Accounting for Climate Change

Accounting for Climate Change

Uncertainty in Greenhouse Gas Inventories – Verification, Compliance, and Trading

Edited by

Daniel Lieberman
ICF International, Washington, DC, USA

Matthias Jonas
International Institute for Applied Systems Analysis, Laxenburg, Austria

Zbigniew Nahorski
Systems Research Institute, Polish Academy of Sciences, Warsaw, Poland

Sten Nilsson
International Institute for Applied Systems Analysis, Laxenburg, Austria

This book would not have been possible without the help of:

Systems Research Institute
Polish Academy of Sciences

 Springer

A C.I.P. Catalogue record for this book is available from the Library of Congress.

ISBN:978-1-4020-5929-2 (HB)
ISBN:978-1-4020-5930-8 (e-book)

Published by Springer,
P.O. Box 17, 3300 AA Dordrecht, The Netherlands.

www.springer.com

Cover image of smokestack © JupiterImages Corporation 2007
Reproduced from Water, Air, & Soil Pollution: Focus, Volume 7, Issues 4–5, 2007

Table of Contents

Water Air Soil Pollut: Focus (2007) 7:421–424
DOI 10.1007/s11267-006-9120-8

Accounting for Climate Change: Introduction

Daniel Lieberman · Matthias Jonas · Wilfried Winiwarter ·
Zbigniew Nahorski · Sten Nilsson

Published online: 3 February 2007

Abstract The assessment of greenhouse gases (GHGs) emitted to and removed from the atmosphere is high on both political and scientific agendas internationally. As increasing international concern and cooperation aim at policy-oriented solutions to the climate change problem, several issues have begun to arise regarding verification and compliance under both proposed and legislated schemes meant to reduce the human-induced global climate impact. The approaches to addressing uncertainty introduced in this article attempt to improve national inventories or to provide a basis for the standardization of inventory estimates to enable comparison of emissions and emission changes across countries. Authors of the accompanying articles use detailed uncertainty analyses to enforce the current structure of the emission trading system and attempt to internalize high levels of uncertainty by tailoring the emissions trading market rules. Assessment of uncertainty can help improve inventories and manage risk. Through recognizing the importance of, identifying and quantifying uncertainties, great strides can be made in the process of Accounting for Climate Change.

Keywords Uncertainty analysis · Greenhouse gas inventories · Kyoto protocol · Emissions trading · Verification and compliance

D. Lieberman (✉)
ICF International,
Washington, DC, USA
e-mail: dlieberman@icfi.com

M. Jonas · S. Nilsson
Forestry Program, International Institute for Applied Systems Analysis,
Laxenburg, Austria

W. Winiwarter
Systems Research, Austrian Research Centers - ARC,
A-1220 Vienna, Austria

Z. Nahorski
Systems Research Institute, Polish Academy of Sciences,
Warsaw, Poland

The assessment of greenhouse gases (GHGs) emitted to and removed from the atmosphere is high on both political and scientific agendas internationally. Under the United Nations Framework Convention on Climate Change (UNFCCC), parties to the Convention have published national GHG inventories, or national communications to the UNFCCC, since the early 1990s. Methods for the proper accounting of human-induced GHG sources and sinks at national scales have been stipulated by institutions such as the Intergovernmental Panel on Climate Change (IPCC) and many countries have been producing national assessments for well over a decade. As increasing international concern and cooperation aim at policy-oriented solutions to the climate change problem, however, several issues have begun to arise regarding verification and compliance under both proposed and legislated schemes intended to reduce the human-induced global climate impact.

Pilot and voluntary GHG emissions trading schemes exist in the United States, United Kingdom, Australia and Europe, and—since January 2005—the European Union has become the world's largest legislated GHG emissions trading market. Common to burgeoning market-oriented GHG reduction schemes both worldwide and global, as well as to national GHG inventory analyses, is the concept of single-point estimates of emissions and emission changes. This accounting method raises a host of crucial questions either directly or indirectly linked to the issue of uncertainty, especially in the context of accounting for emission changes, the central focus of the Kyoto Protocol to the UNFCCC.

The issues of concern at the *International Workshop on Uncertainty in Greenhouse Gas Inventories*, held 24–25 September 2004, in Warsaw, Poland, are rooted in the level of confidence with which national emission assessments can be performed, as well as the management of uncertainty and its role in developing informed policy. Jointly organized by the Systems Research Institute of the Polish Academy of Sciences (http://www.ibspan.waw.pl/) and the Austrian-based International Institute for Applied Systems Analysis (http://www.iiasa.ac.at/), the Workshop covered state-of-the-art research and developments in accounting, verifying, and trading of GHG emissions and provided a multidisciplinary forum for international experts to address the methodological uncertainties underlying these activities. The topics of interest covered national GHG emission inventories, bottom-up versus top-down emission analyses, signal processing and detection, verification and compliance, and emission trading schemes.

Central to current international policy concerns and the present discussion alike is the need for a well-defined role—if, in fact, any role is to be played—of uncertainty analyses in national GHG inventories at the country level, as well as in those falling under the purview of international regulatory schemes. International schemes such as EU emissions trading or that set forth by the Kyoto Protocol—if they are to function as binding agreements—must be able to demonstrate that estimates regarding emission changes are not only measurable but also that they comply with an objective and standard measure that ensures consistent treatment of the uncertainty with which they are associated. It is thus of primary importance to evaluate of the multiple methods through which uncertainty analyses are incorporated into national GHG inventories and the reasons for using them.

While uncertainty estimates are not intended to dispute the validity of national GHG inventory figures, the variability that they communicate underscores the lack of accuracy characterizing many source and sink categories' methodologies and thus makes for a difficult foundation on which to base policy. This does not, however, imply that environmental agencies, corporate environmental departments, and other stakeholders should simply do without uncertainty estimates; on the contrary, a number of arguments illustrate the importance of these analyses.

According to the *IPCC Good Practice Guidance and Uncertainty Management in National Greenhouse Gas Inventories*, uncertainty analysis is intended to help "improve the accuracy of inventories in the future and guide decisions on methodological choice."[1] Uncertainty analyses function as excellent indicators of opportunities for improvement in data measurement, data collection, and calculation methodology; only by identifying elements of high uncertainty can methodological changes be introduced to address them. Currently, most countries that perform uncertainty analyses do so for the express purpose of improving their future estimates; this rationale is generally the same at the corporate level. In either case, estimating uncertainty helps prioritize resources and take precautions against undesirable consequences. Depending upon the intended purpose of an inventory, however, this may or may not be the extent of the utility of uncertainty analysis. Another rationale for performing uncertainty analysis is as a policy tool, a means to adjust inventories and compare emission changes in order to determine compliance. While some authors find the quality of quantitative uncertainty data associated with national inventories insufficient to use for these purposes, a number of studies offer justification for conducting uncertainty analyses to inform and enforce policy

---

[1] J. Penman, D. Kruger, I. Galbally, T. Hiraishi, B. Nyenzi, S. Emmanuel, L. Buendia, R. Hoppaus, T. Martinsen, J. Meijer, K. Miwa and K. Tanabe (eds.) (2000). *Good Practice Guidance and Uncertainty Management in National Greenhouse Gas Inventories.* Institute for Global Environmental Strategies, Hayama, Kanagawa, Japan, p. 6.5. Available at: http://www.ipcc-nggip.iges.or.jp/public/gp/english/.

decisions. Some proposals suggest revising the system of accounting on which current reduction schemes are based, while others seek to incorporate uncertainty measurements into signal analysis procedures that might offer policy makers the advantage of bottom-up/top-down emission verification procedures. Whether uncertainty analysis can or should serve any of these varied purposes, however, continues to be the subject of scientific debate (and an important element reflected in the contents of this volume). The discussions that culminated in this publication—and this volume itself—attempt to bring to light further implications of and rationale for quantifying uncertainty that in many cases have not yet received significant attention from the international scientific or political community.

Single-point emission estimates that do not account for the presence of uncertainty are not likely to be respected by the scientific community as accurate assessments unless many input factors and methodologies undergo some degree of scrutiny beyond that which has been afforded by governments to date. It is generally understood that the current scientific methods used to measure data, as well as those used to calculate emissions, are only accurate within a range, or to a certain degree. It is important to measure and communicate what this degree (of confidence) is in order to encourage confidence in the accepted methods and practices. In the realm of international cooperation and emission reduction efforts, especially where market forces are involved, credibility is very important. This criterion comes into play in determining whether country commitments have been met, and is crucial if comparisons are to be made from one country to the next. Bottom-up versus top-down verification, for example—comparing a traditionally estimated inventory with an alternative inventory that uses atmospheric or remote sensing measurements—offers a significant opportunity through which to improve credibility.

A more regulatory approach suggests using consistent emission estimation algorithms among countries, seeking to minimize the uncertainty inherent in the differences among national estimates by isolating common uncertainties. Another approach introduces the concept of *effective* emission permits. The value of an effective emission permit is determined by the uncertainty associated with the measurements of the emissions that the permit represents.

Consequently, permits' market values increase as their uncertainties decrease. This concept builds upon that of *undershooting*, according to which entities can only prove compliance by reducing emissions to such a level as to minimize the risk of non-compliance (i.e., emissions must be reduced far enough below the target that it can be said, with some degree of confidence, that the target was actually met). Both of these approaches require an accepted reference reduction or detectability level that is valid for all countries; the institutionalization of either of them, however, requires that reliable quantitative uncertainty assessment be incorporated into policy design, which—as noted above—remains a matter of significant discussion.

Through combining emissions studies and economic evaluation, it is possible to compare uncertainty levels in different emissions trading schemes. It has been shown that the boundaries delimited by legislation can significantly influence the credibility of the results (i.e., achievements) of the legislation. For example, significant uncertainty is introduced by a scheme covering all GHGs, such as that introduced by the Kyoto Protocol, compared to the EU emission trading scheme (which currently covers only CO_2). A more rigid emission accounting system than that currently employed by national inventory agencies might allow for country-specific flexibility while ensuring a greater inter-country comparability of emission estimates and their uncertainties.

The approaches to addressing uncertainty discussed in this volume attempt to improve national inventories or to provide a basis for the standardization of inventory estimates to enable comparison of emissions and emission changes across countries. Some seek to use detailed uncertainty analyses to enforce the current structure of the emissions trading system while others attempt to internalize high levels of uncertainty by tailoring the emissions trading market permits. These approaches all agree, however, that uncertainty analysis is a key component of national GHG inventory analyses. The issues that are raised by authors featured in this compilation—and the role that uncertainty analyses play in many of their arguments and/or proposals—highlight the importance of such efforts.

While the IPCC clearly stresses the value of conducting uncertainty analyses and offers guidance on executing them, the arguments in favor of performing these studies go well beyond any suggestions made by the IPCC. Several potential reasons for national GHG inventory teams to continue to improve and standardize the research and estimation methodologies that lead to quantifiable estimates of uncertainty associated with GHG inventories are noted in the text box below. These aims were identified during Workshop discussions, and many are covered in detail by the expanded papers that appear in the following chapters. The papers adhere to the structure of the Workshop. Paper 1 reflects the keynote lecture of the Workshop, while the remaining chapters center on its main three themes: bottom-up and top-down emission inventory approaches (Papers 2–5), signal detection and analysis techniques (Papers 6–7), and the role of uncertainty in emissions trading schemes (Papers 8–11).

Rationale for Improving and Conducting Uncertainty Analyses

1. Only by carrying out research on uncertainties are we better equipped to handle uncertainties in the future.
2. Uncertainty analyses provide a standard measure that can facilitate the process of comparing national GHG inventories one to the other.
3. Uncertainty analyses help to identify the most prudent opportunities for improvement in the methods and estimates of GHG emissions and emission changes in national assessments.
4. Uncertainties play a role in determining whether or not commitments are credibly met.
5. Solid uncertainty assessments have the potential to contribute to the stability of emissions trading markets by reinforcing the value of credible reductions.
6. The Kyoto Protocol will be made more robust in the future by setting targets (ex ante) that explicitly account for the uncertainties associated with emission changes.

Water Air Soil Pollut: Focus (2007) 7:425–441
DOI 10.1007/s11267-006-9119-1

Uncertainties of a Regional Terrestrial Biota Full Carbon Account: A Systems Analysis

S. Nilsson · A. Shvidenko · M. Jonas ·
I. McCallum · A. Thomson · H. Balzter

Received: 5 October 2006 / Accepted: 26 December 2006 / Published online: 30 January 2007
© Springer Science + Business Media B.V. 2007

Abstract We discuss the background and methods for estimating uncertainty in a holistic manner in a regional terrestrial biota Full Carbon Account (FCA) using our experience in generating such an account for vast regions in northern Eurasia (at national and macroregional levels). For such an analysis, it is important to (1) provide a *full* account; (2) consider the relevance of a *verified* account, bearing in mind further transition to a *certified* account; (3) understand that any FCA is a fuzzy system; and (4) understand that a comprehensive assessment of uncertainties requires multiple harmonizing and combining of system constraints from results obtained by different methods. An important result of this analysis is the conclusion that only a relevant integration of inventory, process-based models, and measurements in situ generate sufficient prerequisites for a verified FCA. We show that the use of integrated methodology, at the current level of knowledge, and the system combination of available information, allow a verified FCA for large regions of the northern hemisphere to be made for current periods and for the recent past.

Keywords terrestrial biota · regional full greenhouse account · uncertainty · verification · certification · Northern Eurasia

1 Introduction

From what we know about interactions between the biosphere and the atmosphere, we can assume that only a full carbon account (FCA) (both in itself and as the informational and methodological nucleus of the full greenhouse gas account) corresponds to the essence and ultimate goals of the United Nation Framework Convention on Climate Change (UNFCC) (Nilsson et al., 2000a; Schulze, Valentini, & Sanz, 2002). Because of various political and economic constraints and considerations, the Kyoto Protocol and recent documents of the Intergovernmental Panel on Climate Change (IPCC) still operate with partial carbon accounting systems connected to the managed part of the biosphere. We assume from recent developments that transition to full accounting will be put on the climate change science agenda in the near future.

S. Nilsson (✉) · A. Shvidenko · M. Jonas · I. McCallum
International Institute for Applied Systems Analysis,
2361 Laxenburg, Austria
e-mail: nilsson@iiasa.ac.at

A. Thomson
Center For Ecology and Hydrology,
Monks Wood, UK

H. Balzter
Department of Geography, Climate and Land Surface Systems Interaction Centre (CLASSIC),
University of Leicester, Bennett Building,
University Road, Leicester, LE1 7RH, UK
e-mail: hb91@le.ac.uk
URL: http://www.leicester.ac.uk/geography/staff/academic_balzter.html

Springer

Perhaps the most appropriate way of providing a transition from a partial to a full carbon account is differentiation between "assessment" (i.e., the actual exchange of greenhouse gases between the biosphere and the atmosphere) and "accounting" (i.e., what parts of this exchange are eligible for inclusion in the Kyoto and post-Kyoto accounting mechanisms).

The full carbon account has two parts that differ in terms of their nature and methodology: (1) assessing emissions caused by the anthroposphere (for example, by industry and energy); and (2) quantifying interactions of terrestrial vegetation with other components of the biosphere, in particular, the atmosphere. The share of emissions that each of these two components has in the summarized fluxes of the FCA at the national level may be of the same magnitude (e.g., for Russia, see Nilsson et al., 2003a). The experiences of some countries (European Union member states and the United States) show that the estimated uncertainties of carbon dioxide (CO_2) emissions from fuel combustion are low, as a rule in the ± 2–4% range (confidence level 0.95) (EEA, 2005). In spite of the higher uncertainties for other gases (e.g., in roughly the ± 17–48% range for methane (CH_4) emissions [Monni, Syri, & Savolainen 2004; Rypdal & Winiwarter, 2001]), the overall uncertainties (e.g., expressed on the basis of CO_2 equivalence) of industrial sectors are substantially less than the uncertainties of fluxes resulting from terrestrial vegetation and agriculture (Nilsson et al., 2000a; EEA, 2005). In other words, the uncertainties of the full carbon account will ultimately depend mainly on the uncertainties generated by the biosphere, and the latter is the subject of this analysis.

While the Kyoto Protocol and IPCC documents (IPCC, 1997; 1998; 2000; and 2004b) mention the importance of assessing uncertainty, they do not put this at the center of the problem (e.g., Nilsson, Jonas, & Obersteiner 2000b; Nilsson, Jonas, Obersteiner, & Victor 2001). For instance, the IPCC Guidelines stress that "uncertainty information is not intended to dispute the validity of the inventory estimates, but to help prioritize efforts to improve the accuracy of inventories in the future and guide decisions on methodological choice" (IPCC, 2000: p.6.5). The reliability level of the full carbon account that should be required at the regional and global levels is still being discussed. For the partial account, which is defined by the Kyoto

Protocol and subsequent international documents, Annex 1 countries have a greenhouse gas emission reduction target of 5.2% and the European Union of 8% below 1990 levels by the first commitment period of 2008–2012. This means that the uncertainties for the full carbon account should be minimized to at least a level that is able to provide reliable identification of this reduction. Some scientific discussions (e.g., within the framework of the Global Carbon Project) indicate a presumptive level of ± 20–25% for required limits of uncertainties for summarized continental carbon fluxes (expressed, for example, as net biome production) caused by terrestrial vegetation; this would obviously be too high if the full carbon account were to become a subject of the post-Kyoto negotiation process. Our tentative results for temperate and boreal regions show that FCA uncertainties for large regions could be decreased to a level of ~10–15% (confidence level 0.9); this level at least seems achievable if the FCA meets a number of system requirements and information improvements. The technical jargon, however, requires two clarifications. First, relative errors depend on the estimated mean, and a definite level of uncertainties implies a tacit prerequisite that net biome production (as an eventual estimate of the terrestrial biota full carbon account) is not zero or close to zero. Second, strictly speaking, the completeness of the FCA cannot be estimated in any formal way, and the knowledge and proficiency levels currently available reduce the chances of finding a solution to this problem. Nevertheless, the philosophy behind the FCA does make it possible to develop an approximate solution.

The full carbon account has two major goals that are equally important and interdependent: (1) quantification of all carbon pools and fluxes included in the account; and (2) reliable estimation of uncertainties. The intentions of the UNFCCC and the logic of recent post-Kyoto developments imply the need to move toward a verified full carbon account. A verified account means, following the IPCC, 2000, p. A3.20), that: (1) uncertainties at all stages and for all modules of the accounting scheme are estimated in a comprehensive and transparent way; and (2) the methodology of the FCA should present guidelines as to how uncertainties can be managed, in particular, if the results of the accounting do not satisfy required (preliminary, defined) uncertainty levels. Verification is basically a scientific notion and is (or should be) an inherent part of any accounting scheme.[1] Verification provided by a specially autho-rized independent body could lead to a certified account. Obviously, a certified account should pro-vide a preliminary, defined uncertainty level. Cur-rently, there is no clear understanding as to how it would be possible to build certified systems, partic-ularly at continental and national scales, given the many scientific, political, and institutional problems that would need to be resolved nationally and internationally beforehand (cf. Gillenwater et al., 2007); Jonas and Nilsson (2007); Nahorski and Jęda (2007).

We should not underestimate the difficulties of the transition from the current status through a verified account to a certified account. The Global Carbon Project GCP (2003) indicates that, among inherent shortcomings in quantifying the carbon budget: (1) existing global models are unable to determine carbon sources or sinks with acceptable accuracy at the regional and continental spatial, and interannual time scales; (2) there are no agreements between top-down and bottom-up approaches; (3) there are substantial inconsistencies between global and regional budgets; (4) temporal patterns are poorly understood at timescales greater than a few years; and (5) there are big gaps in our comprehen-sion of the spatial and temporal pattern of human-induced fluxes.

Several methods are used to provide the scientific basis for the terrestrial biota carbon account. The majority of the results at the continental and national levels are received from process-based models and inventory approaches. Each of these methods has well-recognized strengths and weaknesses. During the last decades, advances have been made in process models so that the model structure now explicitly incorporates current knowledge regarding ecosystem processes; process models are practically the only tool available for diagnosing the interannual variation of major carbon fluxes. However, these models operate

with a simplified, mostly "potential" world and do not have an adequate system of uncertainty estimation. While they allow the uncertainties in model projec-tions caused by propagation of uncertainty in model output to be partitioned, they cannot answer the major question of any serious modeling effort, namely, how distant is the model structure from the modeling phenomenon? Attempts to improve uncertainty assess-ment in process-based models (e.g., MacFarlane, Green, & Valentine, 2000; Parysow, 2000; Zaehle, Sitch, Smith, & Hatterman 2005) are limited by intramodel considerations, such as the introduction of variability into input parameters and the assessment of how sensitive model results are to this variability. On the other hand, inventory-based methods, while strong in terms of their empirical basis, are unable to indicate rapid environmental changes or to take into account the temporal trends of major drivers. Other methods used in the FCA, although very important, either serve separate controlling blocks of the account-ing system (inverse modeling) or deliver information for parameterization of the two above background methods (e.g., measuring carbon fluxes in situ).

This paper presents a brief analysis of the experiences and lessons of assessing uncertainties of the terrestrial biota full carbon account at the regional scale for a large region of Siberia through an EU-funded project entitled 'SIBERIA-II' (Multi-sensor Concepts for Greenhouse Gas Accounting of North-ern Eurasia), and from the full carbon account of the entire Russian terrestrial vegetation carried out by IIASA's Forestry Program during recent years. We (1) attempt to illustrate the fact that only a consecutive holistic approach can serve as the background for a verified FCA; (2) briefly analyze the systems require-ments of its structure and methodology; and (3) pres-ent typical examples (see Jonas et al., 1999).

2 Basic Definitions

There are many different approaches to dealing with uncertainty, and selection of the language and dimen-sions involved is of primary importance. We limit our analysis to informational and methodological aspects of the regional FCA, leaving out consideration of the different social, economic, cognitive, institutional, and ethical aspects of the problem.

---

[1] We note that Jonas and Nilsson (2007) go one terminological step further than we do here and strictly distinguish between "validation" and "verification" by applying science-theoretical principles. However, although we use the term "verification" somewhat indifferently, our ultimate understanding of verifica-tion, especially in the context of our integrated (multimeasure-ment/modeling) approach presented here, is in line with the bottom-up/top-down accounting/verification approach discussed by Jonas and Nilsson.

Springer

The terminology used below is generally accepted in statistical theories and risk analysis. The conventional terms for standard statistical analysis are: (1) *precision* as reproducibility or a measure of random error – this deals with our inability to discriminate among values within a parameter or to deal with a parameter's imprecision; (2) *accuracy* as correctness or a measure of the systematic error (bias); and (3) a *mistake* as a measurement that is known to be incorrect due to carelessness, an accident, or the ineptitude of the experimenter. In an FCA, direct use of these terms is usually limited to partial and relatively simple statistical tasks that are based mainly on direct measurements.

Mathematical theory distinguishes between *uncertainty* and *variability*. Although the term *uncertainty* can have different meanings: statistical variability or lack of knowledge, lack of confidence in a single value (Hattis & Burmaster, 1994; Hofman & Hammonds, 1994; Heath & Smith, 2000), its use in global change science is rather consistent. "Uncertainty" is understood as a description of imperfect knowledge of the true value of a particular quantity or its real variability in (1) an individual (e.g., measurements of biometric indicators of trees on a sample plots); or (2) a group (e.g., averages among sample plots established in a homogeneous category of forests). In essence, uncertainty is the absence of information; or it is an expression of the degree to which a value is unknown (IPCC, 2004a; 2004b; Rowe, 1994). Uncertainty can be represented by quantitative measures (e.g., a range of values calculated by various models) or by qualitative statements (e.g., reflecting the judgment of a team of experts). *Variability* is a special contributor to uncertainty. "Interindividual variability" means the real variation within a measured value of individuals or parameters. In general, uncertainty is reducible by collecting additional data or using better models, whereas real variability cannot be changed as a result of better or more extensive measurements. (However, the latter can improve the quality of the estimates used). In our analysis we defined *uncertainty* as an aggregation of insufficiencies of our system output, regardless of whether those insufficiencies result from a lack of knowledge, the intricacies of the system, or other causes (cf. Nilsson et al., 2000a). Finally, uncertainties in the FCA can be expressed as confidence intervals of probability distribution functions.

Probability is the basic term for describing the assessment of any uncertainty. The traditional approach assumes that observed frequencies are equivalent to probabilities – it requires the conditions of the phenomenon or process to remain stationary and for random measurements to take place. However, both these requirements are the exception rather than the rule in an FGA. Moreover, the fuzziness of the FCA inevitably leads to the use of subjective (personal) probabilities, the (FCA-applicable) specifics of which we consider below.

3 Uncertainties of the Regional Full Carbon Account

Strictly speaking, the "ideal" FCA should be the result of continuous monitoring of terrestrial biota in space and time. The philosophy behind this kind of monitoring leads to the idea of an integrated observing system – and beyond, to an integrated accounting system. We can conclude from recent developments that some simplified versions of this type of approach could come to fruition in the near future. Currently, all carbon accounting schemes are forced to use many heterogeneous information sources, including results from different measurements, assessments, and expert estimates over time, which means that numerous and diverse uncertainties are generated. Taking into account the methodological specifics of the carbon account, different classifications (decomposing, categorizing) of uncertainties can be relevant. For the IPCC TAR (Third Assessment Report) Assessment, Moss and Schneider (2000) considered four major groups dealing with (1) confidence in the theory; (2) observations (measurements); (3) models; and (4) consensus within a discipline. Rowe (1994), considering common aspects of risk analysis, divided uncertainties into temporal (past and future), structural (complexity), metric (measurements), and translational (explaining uncertain results). Distinguishing two broad classes of uncertainty-"statistical" (associated with parameter or observational values that are not known precisely) and "structural" (referring to causal relationships between variables)-the IPCC Workshop on Describing Scientific Uncertainties in Climate Change pointed out the substantial difficulties involved in assessing structural uncertainty and the limited opportunities for doing so in any comprehensive and formal way (IPCC, 2004b).

For structuring the FCA uncertainty calculation schemes, a more detailed classification of the sources of uncertainty in the following groups seems useful (see also Jonas et al., 1999; Nilsson et al., 2000a; Shvidenko, Nilsson, Rojkov, & Strakhov 1996):

(1) *Definitions and classification schemes used in calculations.* As a rule, the definitions and classification schemes currently used in the FCA have been introduced for purposes other than carbon accounting and often correspond to inappropriate or obsolete standards and measuring technologies.

(2) *Shortcomings of available data.* Some important data have never been and are not being measured, which leads to incomplete and sometimes inappropriate substitutions.

(3) *Unknown or insufficient precision of measured data.* Reasons for this could vary: for example, subjective (not random) sampling, biased statistics, deliberate falsification, and inappropriate measurement techniques.

(4) *Lack of a proper basis for upscaling.* Very often, there is no solid platform for estimating the accuracy of upscaled point measurements, gradients are unknown, and stratification is provided based on expert judgments.

(5) *Short time series.* Some processes require historical reconstruction for up to 150–200 years, which is not covered by existing historical records.

(6) *Lack of knowledge of some important processes.* For instance, the post-disturbance processes in soil on permafrost, some aspects of below-ground NPP, or nitrogen turnover after biotic disturbances are, to a significant extent, 'black boxes.'

(7) *Oversimplification of the modeling approach.* In both the major methodological approaches of the carbon account (i.e., pool-based and flux-based carbon account), the regional full carbon budget (FCB) is presented by a sophisticated superimposition of (almost exclusively) nonstationary stochastic processes. There is still no methodology that would use this intrinsic feature of an FCB as a prerequisite for its modeling and quantification, and the substitution of deterministic models for stochastic processes is common practice. There are many other examples of this type.

(8) *Spatially and/or temporally insufficient observing systems.* Significant remote areas (e.g., in the Russian north) are not covered by high quality remote sensing (RS) observations (because of the low sun angle and boreal winter night) or by on-ground observations. Some indicators are very dynamic, and existing monitoring systems and available data cannot grasp these dynamics (e.g., seasonal dynamics of insect outbreaks in boreal forests).

Although each class of uncertainties can be addressed separately, the classes are not necessarily independent, and their interdependence should be examined. The above list of uncertainty sources can be applied to some or all periods of the assessment: past, present, and future. However, in any prediction and forecast, many other uncertainties-arising from future drivers (climatic, ecological, social, and economic) and from responses and feedback from terrestrial ecosystems – need to be considered. The level of background uncertainties can be illustrated with reference to the uncertainties of climatic predictions. Using 12 three-dimensional general circulation models (GCMs), including seasonal cycles, a mixed-layer ocean, and interactive clouds and other features, the projected increase in global mean surface air temperature under equilibrium conditions for doubled CO_2 concentrations in the atmosphere varies approximately threefold (from 1.6 to 5.4°C, mean 3.82°C, coefficient of variation 26.3%) (Cess et al., 1993). In spite of obvious progress in climatic modeling during the last decade, the situation has not changed significantly (e.g., Collins et al., 2005). One can conclude that there is no solid background for verified FCA for future time periods. We will not consider this special (and highly uncertain) case further.

Considering the essence, as well as the learning limitations, in terms of information and methodology, of a full carbon budget of terrestrial biota, we can conclude that any FCA is a typical *fuzzy* system. In spite of thousands of publications on this topic since Zadeh (1965) published his fundamental paper, there is no single unique definition of fuzzy systems and/or fuzziness. Thus, we use this term in its rather common but wide mathematical sense (Kosko, 1994;

Wang & Barret, 2003), bearing in mind that many elements of the FCB (procedures, components, and stages of the FCA), far from presenting a crisp set, require the knowledge of multivalued membership functions. In essence, "fuzzy logic is part of a formal mathematical theory for the representation of uncertain systems" (Cogan, 2001); according to Mendoza and Sprouse (1989) "the concept has generally been associated with complexity, vagueness, ambiguity, and imprecision" which "further implies that model coefficients, parameters, or functional relationships may be fuzzy and, hence, not known with complete certainty." The comprehensive development of the formal theory, which would provide for learning about natural fuzzy systems is, to a significant extent, a matter for the future. Although fuzzy logic and fuzzy methods are recommended as a means of incorporating subjective information into different aspects of uncertainty assessment (e.g., Haimes, Barry, & Lambert 1994; Hattis & Burmaster, 1994), their applications in ecology and natural management are limited by numerous diverse, albeit partial, tasks (Bare & Mendoza, 1991; Chen & Mynett, 2003; Mendoza & Sprouse, 1989; U. Özesmi & S. L. Özesmi, 2004; Wan-Xiong, Yi-Min, Zi-Zhen, & Fengxiang 2003 etc.). In the framework of FCA, it is productive to apply "fuzzy thinking," whose philosophical approach is of great help in structuring problems, developing a relevant FCA system, and treating uncertainties. Elements of this approach are being introduced little by little into different parts of global change science. In recent years, the philosophy has also been applied to a "multiple-constraint" approach, where heterogeneous data – for example, measurements of fluxes, remote sensing data, data from different inventories – provide constraints in terms of the models used and in assessing results (e.g., Wang & Barret, 2003). Obviously, "fuzzy thinking" can include a formal definition of membership functions and inference rules, but it is not exhausted by exclusive applications of fuzzy logic methods.

Fuzzy thinking leads to an important conclusion that defines a relevant specific methodology for verified FCA: strictly speaking, no individual FCA method or model applied separately can provide a sufficient (i.e., comprehensive, transparent, and reliable) estimation of uncertainties. Fuzzy thinking thus defines the need to systematically integrate relevant methods and models,

and it leads to the philosophy of *integration* in all its ramifications. For the FCA, the solution is an integration of all relevant information sources (on-ground, remote sensing data, and appropriate regional ecological models), both soft and hard knowledge. On the other hand, integration should be provided for different components of the FCA: carbon of terrestrial biota, ocean, and atmosphere. Consistency in the terrestrial biota global carbon budget is an indicator of its reliability. Comparing the results obtained by different methods is an important part of verification.

The need for *a full* carbon account generates an additional dimension of uncertainty. By definition, "a full C budget encompasses all components of all ecosystems and is applied continuously in time" (Steffen et al., 1998). However, in spite of progress over the last decade, there remain substantial uncertainties in understanding regional and global carbon budgets. This means that the completeness of the FCA can be estimated only through expert judgment.[2] However, estimating an FCB continuously in time in order to judge its completeness can also only be fulfilled in a very approximate manner. As the FCB has a "memory," up-to-date estimates of carbon (C) fluxes may depend strongly upon the previous, sometimes long periods for which relevant measurements may not be provided, and thus the required information simply does not exist. Moreover, the completeness greatly depends upon the end-point target of the user. For example, the final goal of carbon accounts can be defined either as an assessment of the amount of $C–CO_2$ in the exchange, or as the quantities of all gases containing carbon, or as the Global Warming Potential. Nevertheless, experiences of the FCA for some countries (like Austria and Russia) show that about 96–98% of recognized carbon fluxes are usually included in the consideration, although in essence this conclusion is an expert estimate (Nilsson et al., 2000a). The completeness allows us to implement a balance estimation and an analysis of the consistency of individual modules and blocks of the FCA. Here, we face a substantial methodological shortcoming of any partial accounting system: the inability either to close the balance or to

---

[2] We distinguish between a full carbon budget (FCB) as a natural system and a full carbon account (FCA) as an artificial accounting system.

check the consistency of the accounting system. The crucial assumption underlying the partial carbon account is that some drivers and, consequently, some net carbon fluxes (especially those that are not directly human-induced) are untested, and their changes remain unknown. Thus, the FCA presents additional information that allows the (final) uncertainties of the accounting systems to be estimated and the specifics, strengths, and weaknesses of partial accounting systems to be grasped.

4 Requirements for the Terrestrial Biota Regional Full Carbon Account

The above considerations give rise to the following important requirements for any verified FCA result:

1. Only a holistic system approach (with modifications resulting from the fuzziness of the FCA) can serve as a solid overall methodological background for the FCA. From a substantive point of view, implementation of the landscape-ecosystem methodology is one of only a few possibilities for a consecutive system analysis. Under the landscape-ecosystem approach, (1) an ecosystem (i.e., vegetation–soil ensemble at different scales) is considered as the primary unit of scientific description, modeling, and interpretation; and (2) the quantification of intra-ecosystem processes of energy and matter exchange should include the impacts of properties of an individual landscape. From an informational point of view, all relevant sources of information must used, including: (1) as comprehensive a ground-based quantitative descriptions of ecosystems and landscapes as possible (e.g., in the form of Geographic Information Systems); (2) remote sensing data; (3) numerous and diverse sets of auxiliary models (e.g., for connecting remotely sensed data with "hidden" ecological parameters of ecosystems); (4) measurements of fluxes (such as Net Ecosystem Exchange); (5) composition of gas concentrations in the atmosphere; and (6) regional ecological models of different types. From a methodological point of view, a relevant combination of pool-based and flux-based approaches

allows the weaknesses of each of these basic FCA methodologies to some extent to be eliminated.

2. Use of strict and monosemantic definitions and formally complete classification schemes. This problem is not trivial. Recent activities of the Food and Agriculture Organization (FAO) on harmonizing forest-related definitions for use by various stakeholders provide many examples of how many different problems can be encountered in the rather simple field of land use-land cover classifications alone (FAO, 2002).

3. Explicit structuring of the account; use of strict intrasystem (module) spatial, temporal, and process boundaries. In this respect, a number of questions should be regulated (e.g., whether human consumption of vegetation products should be considered as part of net biome production (NBP).

4. An estimation of the uncertainties should be provided at all stages and for all modules of the FCA. This allows the gathering of any additional information needed to understand relevant ways of managing uncertainties.

5. Accounting schemes, models, and assumptions should be presented in an explicit algorithmic form. This means that the use of soft knowledge (e.g., in the form of expert estimates), which is inevitable in the FCA, should be provided in a "quantified" form and using methods that would allow any shortcomings and possible biases resulting from subjective information to be minimized.

6. The accounting scheme should provide a spatially explicit distribution of considered pools and fluxes. This means that all major components of the FCA should be georeferenced at relevant scales.

7. Temporal dimensions of the FCA should be consistent with the temporal peculiarities of processes that are quantified (modeled). The relevant length of respective timescales and the required frequency of observations are defined by specifics of the individual processes considered. Obviously, a year or a different period of accounting should be clearly identified.

Some of the above requirements are not satisfied in regional and national accounting and are not indicated in the recommendations of the IPCC (IPCC, 1997;

2000). This increases the fuzziness as well as the role of expert components in the FCA.

5 Assessing Uncertainties

Two main statistical tools – probability density functions and confidence limits – are normally used for assessing uncertainties. The IPCC Guidelines suggest the use of a 95% confidence interval. This "conventional" recommendation is usually justified in terms of the simplicity of calculating the interval corresponding to two standard errors. It is not completely clear, however, how much this traditional recommendation (1) corresponds to the specifics of the FCA; and (2) impacts interdependence of type I (alpha) and II (beta) errors. Factors that are beyond these two considerations follow.

In essence, the selection of a confidence interval should be based on a function of the losses due to an achieved level of uncertainties. However, not only has no formal theory been developed to quantify such a function but there are large practical difficulties in structuring one. Thus, the solution remains in the field of expert estimates; it should be the result of substantive analysis and, finally, of an agreement among interested parties. Taking into account the utility of the FCA for large territories and the practical consequences of its inherent uncertainties, one might conclude that the relevant confidence interval (e.g., for NBP) should correspond to probability smaller than 0.95 (e.g., in the range of 0.8–0.9) or even smaller. Moreover, such an approach would allow a decrease to a relevant value of errors of type II. This problem, as far as we know, has not been considered in any practical assessment of the uncertainties of the FCA made to date. On the other hand, the numerical expression of uncertainties (i.e., statements such as: "the uncertainty (accuracy) of the final result is at p percent") have a major psychological meaning, at least for the public and policy makers. The inherent uncertainties in some important components of the carbon budget of terrestrial vegetation are high and – if we use a confidence interval for high probability – could be comparable with, or even exceed, 100%. Obviously, any results with uncertainties >100% have no practical meaning. Thus, artificially setting high confidence intervals can give the wrong impression about the practical applicability of the final results of

the FCA. This problem requires further elaboration. In the examples and considerations below we use a confidence level of 0.9.

We examined the following method of assessing uncertainties in the FCA: (1) estimation of precision of all intermediate and final results; (2) "transformation" of precision into uncertainty; and (3) multiple-constraint comparisons of results.

Estimation of precision The FCA is presented as a hierarchical structure of analytical expressions. It allows the formal use of error propagation theory, assuming that the variables used in the calculations are more or less normally distributed. However, only some of the initial data result from direct measurements for which, for example, standard errors and probability distribution functions can be estimated using conventional statistical methods. This generates some open issues: (1) the need to use estimates of the precision of initial variables "by analogy" (i.e., average values by classes of the classification used), or based on expert estimates and subjective probabilities; and (2) the use of "summarized errors" as a substitute for random errors. As a rule, it is impossible to divide the many initial variables used in the FCA into random and systematic errors. Thus, summarized errors are considered functions of both random and systematic errors. In practical situations, the share of bias is relatively small (estimated to be in the order of 10–15% of the random error). In such cases, applying the error propagation theory does not change the essence of statistical conclusions.

"Transformation" of precision into uncertainties The precision calculated is transformed into uncertainty based on sensitivity analysis and expert estimates of unaccounted impacts and processes. The Monte Carlo method is often used as a tool for sensitivity analysis. How this procedure works depends on the end-point target of the assessment.

1. The end point is a fixed but unknown value (e.g., net biome production). Values are sampled at random from distributions representing various "degrees of belief" about the unknown "fixed" values of the parameters (i.e., the true but unknown value is equal to or less than any value selected from distribution). The subjective confidence statement about the true but unknown

assessment end point accounts for multiple sources of uncertainties including, (1) inventory or model structure; (2) presence, variability, and representatives of data; and (3) quantified expert opinions. Uncertainty about a quantity that is fixed (or deterministic) with respect to the assessment end point is often called Type B uncertainty. Variation of input data allows the selection of "important input parameters," which contribute most to the spread in the distribution of the FCA results.

2. The end point is an unknown distribution of values. In such a case, the Monte Carlo simulations are performed in two dimensions producing numerous alternative representations of the true but unknown distributions (assessment of uncertainty of Type A). In practical applications of the FCA, both the above procedures are used; however, it often occurs that a mixture of both types of uncertainties is presented.

Although Monte Carlo calculations are not free from some subjective elements (e.g., a "selection" of the type of unknown distribution), this method presents both comprehensive information about uncertainties of the accounting scheme (model) and important information for management of uncertainties. These results often serve as an iterative step in a process to improve model estimates.

We must note, however, that all these results are true only within the approach (model) used and under given inputs and assumptions; they can have little to do with reality, if the model or assumptions are not "comprehensive" or if they are oversimplified. Thus, if, for example, the model FORCARB (carbon inventory for 2000 for private timberland of United States, which covers about 75% of that country's productive forests) estimates uncertainty as ± 9% of the estimated median of total carbon in the year 2000 and as ± 11% in the projection year 2040 (Heath & Smith, 2000), this just tells us that these results are derived from Monte Carlo calculations within the (rather simple) FORCARB model; they tell us nothing about any "real uncertainty." We have no wish to criticize this particular model – we just use it as an example to demonstrate specifics that are inherent in any model, even the most complicated. Moreover, this explains why an independent and thorough analysis of the completeness and structural rationality of the FCA used is necessary. One

way of providing this analysis is by using expert judgments on the topic; such judgments are quantified and embedded (in addition to Monte Carlo or other methods of sensitivity analysis) in final values of uncertainties (Shvidenko & Nilsson, 2003).

Multiple-constraint comparison of results Three important techniques that allow us to make a final judgment about the FCA are: (1) the balance and consistency analysis of carbon budgets of relatively closed blocks (modules) of the FCA; (2) comparisons of independently calculated intermediate results; and (3) multiple-constraint analysis of final results. We must point out the crucial importance of the multiple-constraint methodology. The "top-down/bottom-up" analysis is currently a major tool for understanding the "real" range of uncertainties of the global carbon budget (see Jonas & Nilsson, 2007). This could be very useful in continental and other macroregional FCAs. Hence, the FCA for Russia shows that the former problem of the missing sink, which has been the subject of intense debate, results from the incompleteness of the account (Nilsson et al., 2003a, 2003b).

The problem of bias A usual prerequisite of uncertainty analysis is that the approaches used do not generate significant bias. As a rule, this assumption is very difficult to check in practical assessments. Bias is often caused by temporal or spatial nonstationarity of processes or of the ways in which measurements are provided. Improving the measurement techniques or methodologies used, as well as integrating new knowledge, could generate a substantial shift in results, indicating previously unrecognized biases. We present two recent examples that illustrate the magnitude of the possible impacts.

1. The first detailed inventory estimate of the net primary production (NPP) of Russian forests for 1993 was based on a database that contained approximately 3,000 sample plots, where measurements were performed by traditional destructive sampling (Nilsson et al., 2000a). These measurements did not account for some of the important components of NPP (e.g., root exudates, which comprise about 15% of the total NPP of boreal forest ecosystems), and they probably underestimated the NPP of fine roots. The transition to a semi-empirical inventory-based modeling system

that does not have significant recognized biases (at the current level of understanding), has increased the average forest NPP in Russia by approximately one-third (Shvidenko, Shepashenko, Nilsson, & Vaganov, 2007). Tendencies of the same magnitude have been also recognized for the NPP of wetlands in Siberia (Vasiliev, Titlyanova, & Velichko, 2001).

2. Based on the remotely sensed normalized difference vegetation index (NDVI), Myneni et al. (2001) have estimated the sequestration of carbon in the above-ground wood of Russian forests to be 283 Tg C yr^{-1} for the period of 1992–1998. This accumulation corresponds to an increase in growing stock volume of about one billion m^3 annually. The forest inventory data for the same period indicate the increase in growing stock to be almost three times less (Shvidenko & Nilsson, 2003). This contradiction has recently been explained (Lapenis, Shvidenko, Sheschenko, Nilsson, & Aiyyer, 2005). The recent analysis of temporal dynamics of the allometric ratios of different phytomass fractions during the last 50 years has recognized the substantially different trends in above-ground wood, green parts, and roots. The calibration procedure provided by Myneni et al. (2001) did not take these dynamics into account. If the findings by Lapenis et al. (2005) are taken into account, the remote sensing estimate is decreasing to a level that is compatible with the forest inventory data.

6 Some Practical Implementations and Results from Case Studies

We attempted to introduce (to the extent possible) the above requirements and techniques while estimating the FCA for two regions with different conditions: (1) Russia as a whole country; and (2) a large (~3 million km^2) region of Northern Eurasia (SIBERIA-II study area, see Box 2). In spite of the availability of information and appropriate levels of detail available for these two subjects being different, the methodology of the two FCAs had many common features. The information base was developed in the form of an integrated land information system (ILIS) which comprises multilayer GIS and corresponding attribute data (at scale 1:2.5 million for the entire country and 1:1 million for the SIBERIA-II region). All relevant information sources were used for the development of

the ILIS: available maps and legends; data from different inventories (in particular, forest inventory) and surveys; various scientific archives; and official statistical data. The landscape-ecosystem methodology served as the overall scientific basis of the account, which was based on an integration of pool-based and flux-based approaches. The flux-based approach is expressed as assessing fluxes (measured in units of carbon per unit of time (e.g., Tg C yr^{-1}) at boundaries of terrestrial ecosystems with other components of the biosphere (atmosphere, lithosphere, hydrosphere)

$$NBP = NPP - HSR - DEC - D - TL - TH, \quad (1)$$

where NBP and NPP are net biome and net primary production; HSR is heterotrophic soil respiration; DEC is flux due to decomposition of coarse woody debris; D is flux due to disturbances; and TL and TH are fluxes to lithosphere and hydrosphere. The pool-based method estimates carbon pools at the beginning and end of the assessment period. A combination of these two approaches (or – in an ideal case – a comparison of independently obtained results) allows us to estimate the methodological consistency of the FCA.

We present some typical examples from the above two case studies. For the whole country we provide our estimation of the FCA for the initial period of the Kyoto Protocol (1988–1992). We must note that the terminology of the Protocol ("since 1990") is not completely appropriate for an independent estimation of any solid carbon budget at the national level, whether full or partial carbon account is considered: information required for an FCA of large territories cannot be made operational at the yearly timescale. Thus, the estimation of uncertainties was provided for 5-year averages (1988–2002).

Two major conclusions follow from the FCA for terrestrial biota of all Russia: (1) the resulting uncertainties in the FCA are relatively high, the net biome production (including human consumption of vegetation products) being estimated as 0.35 ± 0.18 Pg C yr^{-1}; (2) the greatest uncertainty lies in assessing soil processes (the change of soil organic carbon was estimated as -0.04 ± 0.16 Pg C yr^{-1}). Attempts to apply the pool-based method to assessing soil carbon dynamics were insufficient because the requisite information was lacking. It has been shown, however, that major improvements in the reliability of results are possible only in the framework of the full account, and that the problem of the "missing sink" is a problem of the incompleteness of assessments (Nilsson et al., 2003a, 2003b). The overall methodological lesson from this study was that any crucial decrease in uncertainties in the FCA at the national level requires substantial improvements to input information, which could probably be adequately carried out within the framework of integrated observing systems like GEOSS (Global Earth Observation System of Systems). Comprehensive use of (only) existing information could supply satisfactory results in assessing relatively simple components of the FCA, although even in such cases substantial expert elements remain (see Box 1 for assessing uncertainties of forest phytomass, as an example).

Box 1 Uncertainties of estimation of the total amount of phytomass in Russian forests based on forest inventory data

Initial assumptions: Data of the state forest account (SFA; i.e., aggregated data of forest inventory by ~2,000 forest enterprises) and regression equations of phytomass do not have any bias at an accepted level of significance. To check these assumptions, a special statistical and expert analysis of data and procedures has been provided.

Indexes used: i is phytomass fraction, $i=1, ..., 7$; ρ is dominant species, $\rho=1, ..., 27$; m is ecoregion, $m=1, ..., 141$, k-number of forest stands.

Variables: M-mass (dry matter) of fractions, Tg; GS-growing stock volume, m³; A, SI, RS-age, site index, and relative stocking, respectively; δ-content of carbon in phytomass.

Initial data are presented in the form of a matrix for each of the 141 ecoregions across the country; these contain area and growing stock distributed by age classes A for dominant species ρ and types of inventory r ($r=1, 2, 3$), as well as average SI and RS by species and inventory types.

Mass M of fraction i, dominant species ρ, ecoregion m is calculated as

$$M_{i\rho m} = \delta_i \sum_{A=1}^{q} R_{i\rho mA} \cdot GS_{\rho mA}$$

$$= \delta_i \sum_{A=1}^{q} c_0 SI^{c1} A^{c2+c3RS+c4RS2} \qquad \text{(B1.1)}$$

where R is ratio of phytomass fraction to growing stock (expressed as a multidimensional regression of A, SI, and RS) and $c_0, ... c_4$ are regression coefficients.

Thus, the total phytomass of Russian forests is

$$M = \sum_{m=1}^{141} \sum_{\rho=1}^{27} \sum_{i=1}^{7} M_{i\rho m} \qquad \text{(B1.2)}$$

Based on standard methods of error propagation theory, the summarized error of Eq. B1.2 could be expressed in an explicit way (Nilsson et al., 2000a). Applying the set of equations for $R_{i\rho m}$ (Shvidenko, Shepashenko, & Nilsson, 2002), we have the summarized error of forest phytomass by ecoregion in the range of ±5–14% (here and below the confidence level is 0.9) and the final precision (weighted by total mass of phytomass of the ecoregion) is estimated at about ±3%. Assuming the relative error of $\delta_i=\pm2\%$, we come to the final conclusion that the total summarized error is ±3.7%, and the confidence interval is 32.9±1.2 Pg C. This result represents a formal estimate of precision. To assess the extent to which the expert estimates and assumptions used were able to impact this conclusion, five Russian experts were requested to estimate the completeness of the accounting. They unanimously concluded that the assessment accounted for "not less than two-thirds of all uncertainties," that is, the final uncertainty was estimated to be about ± 4.5%. Additional information can be presented by comparison with independent estimates from other sources. However, from nine different estimates of forest phytomass in Russia reported during the last two decades, we were able to select only four that used sufficient information and accepted methodologies. The average densities for forest phytomass in Russia from these sources were: Alexeyev and Birdsey (1994): 3.63 kg C m⁻²; Isaev, Korovin, Utkin, Pryashnikov, and Zamolodchikov (1995): 4.55 kg C m⁻²; Isaev and Korovin (1998): 4.51 kg C m⁻², and IIASA (independent GIS-based method; Nilsson et al., 2000a): 4.403 kg C m⁻². The average of these estimates is 4.27 kg C m⁻², or −0.7% of our estimates of 4.30 kg C m⁻².

Nevertheless, the following considerations illustrate some "hidden" uncertainties that cannot be recognized by any formal analysis. The estimate of 4.30 kg C m⁻² was obtained based on a set of models developed using experimental data available before 1997. The models were recalculated using additional experimental data accumulated in 1997–2004 (the number of sample plots was increased by about 10% to about 3,600). The new set of models was applied to the same data of the (State Forest Account) SFA–

1993, and a new estimate of the density was 4.43 kg C m^{-2}, or about 3% more than the previous one. Obviously, this is within the probabilistic limits of the uncertainty estimated above. Finally, the assumption that the SFA growing stock has no bias is not true: an estimate of the bias for 1993 is +2.5% (Shvidenko & Nilsson, 2002). However, 59% of Russian forests (by growing stock) were composed of mature, overmature, and uneven-aged forests with a substantial amount of trunk and root decay. An approximate conservative estimate gives the values as 2–3% of the growing stock (i.e., we have an approximate compensation of the bias of the growing stock estimation). Thus, the overall conclusion is that the uncertainties inherent in our knowledge of the phytomass of Russian forest ecosystems in 1993 are at the level of 5–6% with high probability (not less than 0.9).

SIBERIA-II aimed to make a full greenhouse gas account based on a fusion of (1) multisensor remote sensing; (2) comprehensive description of individual ecosystems and landscapes in the form of an ILIS; and (3) use of different types of ecological model. SIBERIA-II had a number of features that helped substantially increase the reliability of the regional FCA. First, the introduction of multisensor remote sensing greatly increased the quality and efficiency of information. Considering the large scale and remoteness of the region, the information presented by RS (12 different sensors were examined) was of crucial importance for, inter alia, updating land cover, estimating disturbances, and assessing environmental indicators. However, there were many inconsistencies in the technical capacities of RS sensors, the spatial and temporal resolution needed, and the requirements of the FCA. There was an obvious need for new technical RS tools designed specially for studying the biospheric role of terrestrial biota, a good example being a satellite with P-band radar on board for assessing vegetation (particularly forest above-ground biomass). Second, the objective in using diverse information was to increase the synergy from combining the various relevant information sources. Third, applying different ecological models presented the possibility not only of multiple constraints of the results but also of independent estimates of many components of the FCA.

The examples presented below are typical. They are limited by the approach, which is based on the ecosystem-landscape methodology. An FCA-relevant GIS layer and corresponding databases were developed at the polygon level. An FCA is provided for each of the polygons (which serve as a primary ecosystem landscape unit and are aggregated into ecoregions). Some of the components of the FCA are estimated based on regional ecosystem-landscape models. This puts special requirements on the hierarchical structure of the classification of land classes used to limit the variability of the FCA components within the classes. From a modeling point of view, the approach consecutively examines three FCA varieties: (1) "baseline" inventory, assessing average values; (2) introduction of a number of environmental indicators by using empirical and semi-empirical ecosystem and landscape models; and (3) use of process-based blocks as part of the multiple-constraints procedure.

The most important lessons learned from this regional case study are:

1. The study has supported the appropriateness of an ecosystem-landscape approach as the scientific background for the regional FCA.
2. The vegetation components of the FCA for individual polygons are estimated with high reliability. Hence, live biomass (phytomass) by polygons is defined with uncertainties ±7–15%, net primary production and heterotrophic soil respiration ±15–20% (confidence probability here and below 0.9). However, this aspect required the development of a number of special regional modeling systems based on a large number of sample plots (between several hundred and several thousand for each component) and the use of all available reference and normative information (e.g., yield tables and models of gross and net growth).
3. The uncertainty of estimates of soil carbon pools is high (in the range of ±10–15%) and contains a substantial share of expert elements and assumptions because of the coarse resolution of soil data (the basic soil map and reference databases are presented at a scale of 1:1 million), obsolete and unevenly distributed measurements, mapping at different time scales, and insufficiently documented history of vegetation fire during the last two decades. At the ecoregion level, uncertainties of major pools and fluxes (like NPP and HSR) are estimated to be in the range of 5–10% (each ecoregion contains 600–4,000 polygons), under the

assumption that the account has no significant bias (more information and typical examples are given in Box 2). Calculations provided by both pool-based and flux-based methods showed rather consistent results, although assessing the soil carbon dynamic is substantially less certain than for other carbon pools (phytomass, coarse woody debris).

4. Some problems with estimating uncertainties are generated by the aggregation of ecosystems in polygons taking into account the coarse scale of the accounting. To some extent these uncertainties are decreased by the implementation of "mixed classes" (e.g., polygons that contain more than one class). On the other hand, implementation of "virtual polygons" presents the additional possibility of decreasing uncertainties of this type. "Virtual polygons" comprise land classes that are represented by numerous plots of small areas and are not individually indicated at the GIS layer (roads, small rivers and water reservoirs, settlements, some classes of agricultural lands). As a rule, the total area of such land classes could be obtained from independent sources, and corresponding corrections of an area are provided at the ecoregion level. However, a substantial part of the aggregation is based on professional judgments, and estimating these uncertainties includes a substantial expert component.

5. Interannual variability of the FCA could be very high (up to 2–5 fold for NBP and up to 25–30% for NPP during a 10–15 year period) and is defined by the impacts of seasonal weather specifics and by the extent and severity of disturbances.

6. Uncertainties of an FCA estimated for an individual year could be very high. Thus, considering time series is the best strategy for reducing uncertainty.

Box 2 Monte Carlo estimation of uncertainties of phytomass, NPP, and net ecosystem production (NEP) at the regional level (SIBERIA-II) for a base year 2003.

The region of SIBERIA-II, a total area of 307.8 million ha, stretches for about 3,000 km from the Arctic Ocean to the boundary with the Tuva Republic in the south and includes the main vegetation zones of the northern hemisphere (polar desert, tundra, forest tundra, northern, sparse, middle and southern taiga,

temperate forests, forest steppe, steppe and semidesert). The area of the region is divided in 23 ecological regions (ecoregions) and ~35,000 polygons of which 16,589 are covered by vegetation (ecosystem-landscape units). The FCA was provided by polygon. Phytomass by seven fractions was estimated as described in Box 1. Net Primary Production (NPP) was calculated based on a special method of modeling of the annual cycle of total production of phytomass (TPPh). The method, algorithm, and parameterization used are described in Shvidenko, Shepashenko, Nilsson, and Bouloui (2004). The estimation of the FCA was provided similarly to Eq. 1 with some technical modification. Monte Carlo simulations (15,000 runs per simulation) were provided for phytomass by fractions, NPP, and NEP at both polygon and ecoregion levels. Input uncertainties for simulation were estimated as follows: growing stock ±15–20% (requirements of forest inventory manual addressed to separate stand are ±12–15%), site index ±5%, age ±10–40 years, depending on the average age of stand and the dominant species, relative stocking ±15–20%. We present results of the simulations below.

Estimation of phytomass at the polygon level. For a typical ecoregion (no. 2,501 situated in middle taiga subzone of Irkutsk *oblast*), the uncertainty of the total phytomass varies between ±6% and ±14% (mean 12%). The range of uncertainty is similar for all forest fractions (±13–20%) apart from understory and green forest floor, which have lower mean uncertainties, and foliage, which has a higher upper limit (±21–25%). The size of the 90% confidence interval normalized by area ranges between 0.67 and 31.60 Mg C ha^{-1} (mean 13.44). The spatial distribution of the uncertainties is presented in Fig. 1. As a whole, there are no spatial trends of the magnitude of uncertainty. Confidence intervals are mainly influenced by the average density of total phytomass by unit area.

Estimation of NPP at the polygon level. NPP of three aggregated fractions (above-ground wood, green parts, and below-ground wood) was considered. The range of uncertainty is similar for each forest fraction, from less than ±1% to between ±12 and 14%. The mean values are ±12% for the tree fractions and ±8–9% for understory and green forest floor. Percentiles of 5 and 95% boundaries are similar for the abovementioned fractions, which are on average 0.92 and 1.08 of the mean, respectively.

Fig. 1 Uncertainties of total phytomass by polygon for ecoregion 2501. The location of the ecoregion in the SIBERIA-II region is shown in Fig. 2

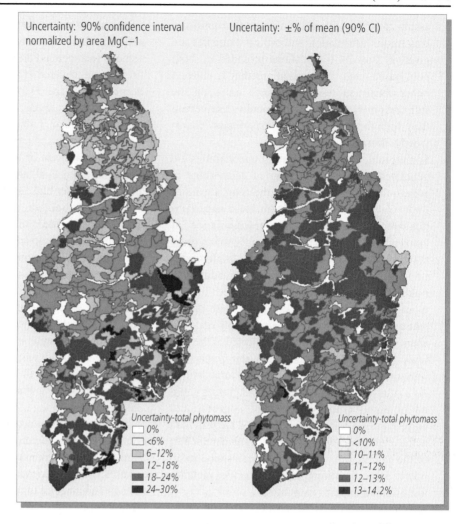

Estimation of NEP. An estimation of uncertainties of assessment of NEP has been carried out at the polygon level for each ecoregion and for the region of SIBERIA-II as a whole. The normalized range of NEP in ecoregion 2,501 varies between 0.17 and 0.85 Mg C ha^{-1} (mean of 0.67 Mg C ha^{-1}). The normalized range of NEP varies across the Siberia-II region between 0.01 and 2.64 Mg C ha^{-1} (mean of 0.51 Mg C ha^{-1}). There is a clear spatial trend in distribution of uncertainties across the region's area (Fig. 2) which is explained by the increasing human impact on ecosystems from north to south.

There are different ways of managing uncertainties based on additional information. There are many ways of evaluating the value of information, most of which rely on determining the benefit of making a decision based on current knowledge versus spending

more resources to improve the knowledge base that could be used in Bayesian decision analysis (Berger, 1985), or referring to the more familiar expected value of perfect information (Morgan & Henrion, 1990). Effective ways of reducing carbon flux uncertainties strictly depend on the structure and specifics of the accounting schemes; the most appropriate ways of reducing their uncertainties differ from those used to reduce uncertainties in the inventories of carbon pools. As a rule, an optimal way of reducing uncertainty requires a systems approach and lies in the attempt to utilize the synergism of combining heterogeneous information sources. For example, to substantially reduce the uncertainties of emissions caused by vegetation fires, more appropriate classifications (for example, types of fires, types of combustibles) are required than are used in many countries; also required are more accurate vegetation

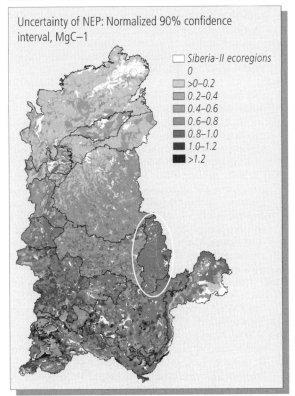

Fig. 2 Uncertainty of NEP for all vegetation classes across the Siberia-II region (uncertainty is shown as the 90% confidence interval normalized by polygon area). The red ellipse identifies ecoregion 2501 depicted in Fig. 1

fuel maps, new or modified RS sensors (which enable types of fires and their severity to be identified), and improved empirical models (e.g., to assess the amount of consumed combustibles of definite forest types depending on such factors as environmental indicators, and fuel storage). In addition, it must be kept in mind that some uncertainties cannot be reduced, given current knowledge and economic conditions.

7 Conclusion

The development of global integrated observing systems is a major strategy that aims to establish verified regional terrestrial biota full carbon accounts in the future. The integrated observation system is understood as a permanent tool for combining all the relevant information sources (on-ground measurements, remotely sensed data and empirical knowledge) and models of

different types linked to primary polygons relevant to the FCA. Some prototypes for components of such systems and possible decisions are outlined above. Presumably, such an approach would allow the uncertainties of annual NBP at regional and national scales to be decreased to a range of 7–10%. However, any proper development and implementation of such a system will require not only advanced theoretical and technical improvements but also the development of new elements and subsystems. These improvements mostly deal with remote sensing, the study of some poorly understood basic processes, and the development of new types of regional model. Remotely sensed data are vitally important for the FCA. However, (1) only a multisensor remote sensing concept will be able to satisfy the major requirements of the accounting; and (2) there is an obvious need for the development of new sensors that would specifically address the assessment of the basic components of the FCA. One of the main bottlenecks of the FCA is insufficient knowledge of ecosystem below-ground processes. From the modeling point of view, it is clear that results produced by dynamic global vegetation models (DGVMs) for individual countries and continents have little in common with reality and that their uncertainties still cannot be estimated in any formal way. Inventory-based modeling schemes are able to present only average data for a rather uncertain period. Recent developments show that the "regionalizing" of DGVMs is one way of introducing such models into the verified FCA (Beer, Lucht, Schmullius, & Shvidenko, 2006). In addition, there are promising results from the introduction of process-based elements in inventory-based approaches and the ways in which this was carried out under the SIBERIA-II project. These can be considered as steps toward developing new types of hybrid regional model which would keep the strengths and minimize the weaknesses of both inventory- and process-based approaches.

One important unresolved question is the setting of the thresholds of relevant uncertainties that should be provided by verified regional and national FCAs. There is little progress in this field to date.

In the foreseeable future, the FCA will remain a fuzzy system in the sense discussed above. This implies that judgments about the reliability of the FCA will be based on a combination of strict formal methods as well as expert conclusions. In February 2005 the Kyoto Protocol entered into force and the technical task of assessing uncertainties gained political and economic importance. The theoretical and practical aspects of the problem will thus need to be elaborated, and the special institutions that would be responsible for certifying FCAs to be developed.

Acknowledgment The SIBERIA-II project (EVG-2001-00008), 2003–2005, was funded by the European Commission (Generic Activity 7.2: Development of Generic Earth Observation Technologies).

References

Alexeyev, V. A., & Birdsey, R. A. (Eds.) (1994). *Carbon in Ecosystems of Forests and Wetlands of Russia,* Sukachev Institute of Forestry, Krasnoyarsk [in Russian].

Bare, B. B., & Mendoza, G. A. (1991). Timber harvest scheduling in a fuzzy decision environment. *Canadian Journal of Forest Research, 22,* 423–428.

Beer, C., Lucht, W., Schmullius, C., & Shvidenko, A. (2006). Small net carbon dioxide uptake by Russian forests. *Geophysical Research Letters, 33* L15403, doi:10.1029/2006GL0026919.

Berger, J. (1985). *Statistical decision theory & bayesian analysis,* second edition. New York, USA: Wiley.

Cess, R. D., Zhang, M.-H., Potter, G. L., Barker, H. W., Colman, R. A., Dazlich, D. A., et al. (1993). Uncertainties in carbon dioxide radiative forcing in atmospheric general circulation models. *Science, 262,* 1252–1255.

Chen, Q., & Mynett, A. E. (2003). Integration of data mining techniques and heuristic knowledge in fuzzy logic modeling of eutrophication in Taihu Lake. *Ecological Modelling, 162,* (1–2), 55–67.

Cogan, B. (2001). Certainty and uncertainty in science. *Scientific Computing World,* pp. 28–30. (December)

Collins, W. D., Ramaswamy, V., Schwarzkopf, M. D. Y., Sun, R., Portmann, W., Fu, Q., et al. (2005) Radiative forcing by well mixed greenhouse gases: Estimates from climate models in the IPCC AR 4. *Journal of Geophysical Research.* Available at http://www.cgd.ucar.edu/cms/wcollins/papers/.

EEA (2005). Annual European community greenhouse gas inventory 1990–2003 and inventory report 2005. Submission of the UNFCCC Secretariat, revised final version, 27 May, *Technical Report 4/2005* of the European Environment Agency.

FAO (2002–2005). *Proceedings, expertmeetings on harmonizing forest-related definitions for use by different stakeholders: First meeting,* 23–25 January 2002, Rome; *second meeting,* 11–13 September 2002, Rome; *third meeting,* 17–20 January 2005, Rome.

Gillenwater, M., Sussman, F., & Cohen, J. (2007). Practical policy applications of uncertainty analysis for national greenhouse gas inventories. *Water, Air and Soil Pollution:* Focus (in press) doi:10.1007/s11267-006-9118-2.

GCP (2003). Global Carbon Project 2003 Science framework and implementation, Earth System Science Partnership IGBP, IHDP, WCRP, DIVERSITAS, *Global Carbon Project Report No.1,* Canberra, Australia.

Haimes, Y. Y., Barry, T., & Lambert, J. H. (1994). Proceedings of the workshop, 'Where and how can you specify a probability distribution when you don't know much?' *Risk Analysis, 14*(5), 661–706.

Hattis, D., & Burmaster, D. E. (1994). Assessment of variability and uncertainty distributions for practical risk analysis. *Risk Analysis 14,* 713–730.

Heath, L. S., & Smith, J. E. (2000). An assessment of uncertainty in forest carbon budget projections. *Environmental Science and Policy, 3,* 73–82.

Hofman, F. O., & Hammonds, J. S. (1994). Propagation of uncertainty in risk assessments: The need to distinguish between uncertainty due to lack of knowledge and uncertainty due to variability. *Risk Analysis, 14,* 707–712.

IPCC (1997). *Revised 1996 IPCC guidelines for national greenhouse gas inventories. volume 1: Greenhouse gas inventory reporting instructions, volume 2: Greenhouse gas inventory workbook, volume 3: Greenhouse gas inventory reference manual.* IPCC/OECD/IEA. Intergovernmental panel on climate change IPCC working group i wG i technical support unit, Bracknell, United Kingdom. Available at: http://www.ipcc-nggip.iges.or.jp/public/gl/invs1.htm.

IPCC (1998). *Managing uncertainty in national greenhouse gas inventories.* Report of the meeting of the IPCC/OECD/IEA programme on national greenhouse gas inventories, held 13–15 October in Paris, France.

IPCC (2000). Good practice guidance and uncertainty management in national greenhouse gas inventories. In J. Penman, D. Kruger, I. Galbally, T. Hiraishi, B. Nyenzi, S. Emmanuel, L. Buendia, R. Hoppaus, T. Martinsen, J. Meijer, K. Miwa, & K. Tanabe (eds.), *Intergovernmental panel on climate change IPCC national gas inventories program, technical support unit.* Institute for Global Environmental Strategies, Hayama, Kanagawa, Japan.

IPCC (2004a). *Documents in support of the writing process for the IPCC working group II fourth assessment report.* Volume produced for the first Lead Authors' Meeting, held 20–23 September in Vienna, Austria.

IPCC (2004b). Describing scientific uncertainties in climate change to support analysis of risk and of options. In M. Manning, M. Petit, D. Easterling, J. Murphy, A. Patwardhan, H.-H. Rogner, R. Swart, & G. Yohe (Eds.), *Report on IPCC Workshop,* held 11–13 May in Maynooth, Co. Kildare, Ireland. Available at http://ipcc-wg1.ucar.edu/meeting/URW/.

Isaev, A. S., & Korovin, G. N. (1998). Carbon in forests of northern Eurasia. In G. A. Zavarzin (Ed.), *Carbon turnover in territories of Russia* (pp. 63–95). Moscow: Ministry of Science and Technology of the Russian Federation, [in Russian].

Isaev, A. S., Korovin, G. N., Utkin, A. I., Pryashnikov, A. A., & Zamolodchikov D. G. (1995). Carbon stock and deposition in phytomass of the Russian forests. *Water, Air and Soil Pollution, 70,* 247–256.

Jonas, M., & Nilsson, S. (2007). Prior to economic treatment of emissions and their uncertainties under the Kyoto Protocol: Scientific uncertainties that must be kept in mind. *Water, Air and Soil Pollution:* Focus (in press.) doi:10.1007/s11267-006-9113-7.

Jonas, M., Nilsson, S., Shvidenko, A., Stolbovoi, V., Gluck, M., Obersteiner, M., et al. (1999). *Full Carbon Accounting and the Kyoto Protocol: A Systems-Analytical View,* Interim Report IR-99-025, International Institute for Applied Systems

Analysis, Laxenburg, Austria. Available at: http://www.iiasa.ac.at/Publications/Documents/IR-99-025.pdf.

Kosko, B. (1994). *Fuzzy thinking*. London, UK: Flamingo.

Lapenis, A., Shvidenko, A., Sheschenko, A. D., Nilsson S., & Aiyyer A. (2005). Acclimation of Russian forests to recent changes in climate. *Global Change Biology, 11*, 1–13.

MacFarlane, D. W., Green, E. J., & Valentine, H. T. (2000). Incorporating uncertainty into the parameters of a forest process model. *Ecological Modelling*, (1): 27–40.

Mendoza, G. A., & Sprouse, W. L. (1989). Forest planning and decision making under fuzzy environments: An overview and illustration. *Forest Science, 32*, 481–502.

Monni, S., Syri, S., & Savolainen, I. (2004). Uncertainties in the finnish greenhouse gas emission inventory. *Environmental Science and Policy, 7*, 87–98.

Morgan, M. G., & Henrion, M. (1990). *Uncertainty: A guide to dealing with uncertainty in quantitative risk and policy analysis*. New York, USA: Cambridge University Press.

Moss, R. H., & Schneider, S. H. (2000). Uncertainties in the IPCC TAR: Recommendations to lead authors for more consistent assessment and reporting. In R. Pachauri, T. Taniguchi, & K. Tanaka (Eds.), *Guidance papers on the cross cutting issues of the third assessment report of the IPCCC intergovernmental panel on climate change* (33–51). Geneva, Switzerland.

Myneni, R. B., Dong, J., Tucker, C. J., Kaufmann, R. K., Kauppi, P. E., Liski, J., et al. (2001). *A large carbon sink in the woody biomass of northern forests*. Proceedings of the National Academy of Sciences, 9826, 14784–14789, Washington, D.C., USA: National Academy of Sciences.

Nahorski, Z., & Jęda, W. (2007). Processing national CO_2 inventory emissions data and their total uncertainty estimates. *Water, Air and Soil Pollution:* Focus (in press) doi:10.1007/s11267-006-9114-6.

National Assessment Synthesis Team (2001). Climate change impacts on the United States: The potential consequences of climate variability and change. *Report for the US Global Change Research Program*, Cambridge, UK: Cambridge University Press.

Nilsson, S., Jonas, M., & Obersteiner, M. (2000b). *The forgotten obligations in the Kyoto negotiations,* Document made available on the Internet by the International Institute for Applied Systems Analysis, Laxenburg, Austria, http://www.iiasa.ac.at/Research/FOR/carb_kyoto.html?sb=10.

Nilsson, S., Jonas, M., Obersteiner, M., & Victor, D. (2001). Verification: The gorilla in the struggle to slow global warming. *Forestry Chronicle, 77*, 475–478.

Nilsson, S., Jonas, M., Stolbovoi, V., Shvidenko, A., Obersteiner, M., & McCallum I. (2003b). The missing sink. *Forestry Chronicle, 79*(6), 1071–1074.

Nilsson, S., Shvidenko, A., Stolbovoi, V., Gluck, M., Jonas, M., & Obersteiner, M. (2000a). Full carbon account for Russia. *Interim Report IR-00-021*, International Institute for Applied Systems Analysis, Laxenburg, Austria. Available at: http://www.iiasa.ac.at/Publications/Documents/IR-00-021.pdf, Study also featured in: *New Scientist*, 2253, 18–19, 26 August.

Nilsson, S., Vaganov, E. A., Rozhkov, V. A., Shvidenko, A. Z., Stolbovoi, V. S., McCallum, I., et al. (2003a). Greenhouse gas balance and mitigation strategies for Russia. *Paper given at the World Climate Conference*, held in Moscow, Russia, 29 September–3 October, (Abstracts, 242–243).

Özesmi, U., & Özesmi, S. L. (2004). Ecological models based on people's knowledge: A multi-step fuzzy cognitive mapping approach. *Ecological Modelling, 176*(1–2), 43–64.

Parysow, P., Gertner, G., & Westervelt, J. (2000). Efficient approximation for building error budgets for process models. *Ecological Modelling, 135*(2–3), 111–125.

Rowe, W. D. (1994). Understanding uncertainty. *Risk Analysis, 14*(5), 743–750.

Rypdal, K. L., & Winiwarter, W. (2001). Uncertainties in greenhouse gas inventories – evaluation, comparability and implications. *Environmental Science Policy, 4*, 107–116.

Schulze, E.-D., Valentini, R., & Sanz, M.-J. (2002). The long way from Kyoto to Marrakesh: Implications of the Kyoto Protocol negotiations for global ecology. *Global Change Ecology, 8*, 505–518.

Shvidenko, A., & Nilsson, S. (2002). Dynamics of Russian forests and the carbon budget in 1961–1998: An assessment based on long-term forest inventory data. *Climatic Change, 55*, 5–37.

Shvidenko, A., & Nilsson, S. (2003). A synthesis of the impact of Russian forests on the global carbon budget for 1961–1968. *Tellus, 55B*, 391–415.

Shvidenko, A., Nilsson, S., Rojkov, V., & Strakhov, V. (1996). Carbon budget of the Russian boreal forests: A system analysis approach to uncertainty. In M. J. Apps & D. T. Price (Eds.), *Forest ecosystems, forest management and the global carbon cycle*, (145–162)NATO ASI, Series 1, Vol. 40.

Shvidenko, A., Shepashenko, D., & Nilsson, S. (2002). Aggregated models of phytomass of major forest forming species of Russia. *Forest Inventory and Management, 1*, 50–57, Krasnoyarsk [in Russian].

Shvidenko, A., Shepashenko, D., Nilsson, S., & Bouloui, Yu. (2004). The system of models of biological productivity of Russian forests. *Forestry and Forest Management, 2*, 40–44 [in Russian].

Shvidenko, A., Shepaschenko, D., Nilsson, S., & Vaganov, E. (2007). Dynamics of phytomass and net primary production of Russian forests: New estimates. *Doklady* of the Russian Academy of Sciences (in press).

Steffen, W., Noble, I., Canadell, J., Apps, M., Schulze, E.-D., Jarvis, P. G., et al. (1998). The terrestrial carbon cycle: Implications for the Kyoto Protocol. *Science, 280*, 1393–1394.

Vasiliev, S. V., Titlyanova, A. A., & Velichko, A. A. (eds). (2001). *West Siberian peatlands and carbon cycle: Past and present*, Proceedings of the International Symposium, held 18–22 August at Noyabrsk, Russia.

Wang, Y. P., & Barret, D. J. (2003). Estimating regional terrestrial carbon fluxes for the Australian continent using a multiple-constraint approach: 1. Using remotely sensed data and ecological observation of net primary production. *Tellus, 55B*, 270–279.

Wan-Xiong, W., Yi-Min, L., Zi-Zhen, L., & Fengxiang, Y. (2003). A fuzzy description of some ecological concepts. *Ecological Modelling, 169*(2–3), 361–366.

Zadeh, L. (1965). Fuzzy sets. *Information and Control, 8*, 338–353.

Zaehle, S., Sitch, S., Smith, B., & Hatterman, F. (2005). Effects of parameter uncertainties on the modeling of terrestrial biosphere dynamics. *Global Biogeochemical Cycles, 19*, GB3020.

Water Air Soil Pollut: Focus (2007) 7:443–450
DOI 10.1007/s11267-006-9117-3

National Greenhouse Gas Inventories: Understanding Uncertainties versus Potential for Improving Reliability

W. Winiwarter

Received: 12 May 2005 / Accepted: 24 February 2006 / Published online: 19 January 2007
© Springer Science + Business Media B.V. 2007

Abstract We investigated the Austrian national greenhouse gas emission inventory to review the reliability and usability of such inventories. The overall uncertainty of the inventory (95% confidence interval) is just over 10% of total emissions, with nitrous oxide (N_2O) from soils clearly providing the largest impact. Trend uncertainty – the difference between 2 years – is only about five percentage points, as important sources like soil N_2O are not expected to show different behavior between the years and thus exhibit a high covariance. The result is very typical for industrialized countries – subjective decisions by individuals during uncertainty assessment are responsible for most of the discrepancies among countries. Thus, uncertainty assessment cannot help to evaluate whether emission targets have been met. Instead, a more rigid emission accounting system that allows little individual flexibility is proposed to provide harmonized evaluation uninfluenced by the respective targets. Such an accounting system may increase uncertainty in terms of greenhouse gas fluxes to the atmosphere. More importantly, however, it will decrease uncertainty in intercountry comparisons and thus allow for fair burden sharing. Setting of post-Kyoto emission targets will require the independent evaluation of achievements. This can partly be achieved by the validation of emission inventories and thorough uncertainty assessment.

Keywords model uncertainty · Monte Carlo simulation · greenhouse gases · inventory quality considerations · Kyoto Protocol

1 Introduction

Emission inventories are important tools for environmental policy. Typically covering material flows into the atmosphere, fluxes of atmospherically active substances (air pollutants or greenhouse gases [GHGs]) are accounted for as annual totals for specified regions. In general, the estimation of emissions follows guidelines that leave freedom for country-specific refinements (EEA 2004; Houghton et al. 1997). As direct measurements of emissions are rarely performed, the assessment of emissions from a single source is often based on the multiplication of a statistical parameter "activity" and the relation of this parameter to the emissions – the "emission factor." Even more complex variants of emission calculations (emission models) may be traced back to this simple concept (see, e.g., Webber and Fleming 2002). Greenhouse gas inventories, which also consider sinks of gases, do not explicitly refer to emissions or sources. The full amount of mass transfer into and from the atmosphere is not considered in all cases in

W. Winiwarter (✉)
Systems Research, Austrian Research Centers-ARC,
A-1220 Vienna, Austria
e-mail: wilfried.winiwarter@arcs.ac.at

W. Winiwarter
Atmospheric Chemistry and Economic Development,
International Institute for Applied Systems Analysis,
A-2361 Laxenburg, Austria

🖄 Springer

the inventory: international obligations to report national emissions fail to cover emissions from sources that are not considered attributable to a single nation. For example, international transport and natural emissions are not included in the national totals (UNFCCC 2004).

Increasing regulatory demands require improvements to be made to inventory quality. When a well-defined relationship exists between emissions (from a source) and the impact (on a receptor) of pollution, an emissions estimate provides a sufficient basis for regulatory action. Current research and policy issues of multicompound atmospheric chemistry, transboundary aspects of air pollution, or emissions trading require a much more intrinsic understanding of both the source–receptor relationship and the emission-generating processes. Consequently, efforts to improve emission inventories and to validate inventory output have been initiated, including an assessment of the reliability and uncertainty of inventories as part of these efforts (Penman et al. 2000).

The uncertainty of national greenhouse gas emissions as one component of quality improvement has been assessed for a number of countries (e.g., Monni et al. 2004; Winiwarter and Rypdal 2001). Using the experience gained from these studies, and interpreting the sensitivities associated with the quantitative assessment of uncertainties, a number of important conclusions can be drawn, including consequences for environmental policy.

2 Methodology: How to Assess the Uncertainty of National Emission Inventories

The assessment of the quality of any model result may take one of two different pathways:

1. Independent validation allows an unbiased assessment of model performance.
2. Sensitivity analysis, determining the uncertainty range of model input information and extrapolating to the variability of the output, is possible without independent information.

Because of the lack of independent validation data (such as emission estimates from inverse modeling of measured atmospheric concentrations), the quality of emission inventories can be fully covered at present only by investigating their input data.

In a study accompanying the official Austrian greenhouse gas emission inventory, all input information was systematically evaluated for its uncertainty (Winiwarter and Rypdal 2001). Both the magnitude and shape of the probability density functions were assessed using discrepancies between statistical data, measurements, or literature information as the main sources. Nevertheless, for a number of parameters no such reliable data was available. Structured interviews with experts in the respective sectors were used to obtain a well-documented expert estimate of the uncertainty elements of those parameters.

While this approach is fully able to cover the random variability of the underlying information, a potential systematic error (e.g., caused by methodological limitations) will not be detected. Such an error would, by its very nature, require correction at the time it is discovered; thus, it would not contribute to variability. In the above-mentioned study, systematic errors are assessed by not correcting data that is clearly erroneous. The difference between systematically wrong results and the new estimates is considered to represent systematic errors, assuming that any systematic error still remaining unidentified would be of the same size as those actually discovered.

The combination of uncertainties can be performed by error propagation or by Monte Carlo methods. While the application of error propagation has some theoretical limitations, the Monte Carlo approach requires more computing power, as it is based on random variations of the input parameters according to their respective probability density and on a statistical evaluation of the output. Considering the fairly simple computations involved in emission calculation, computing time is not a real issue, and the advantages of a Monte Carlo simulation, especially in terms of the treatment of covariance between two parameters (see below), become obvious. Sensitivity analysis demonstrates that – independently of the shape of the input probability functions – the output will approximate a normal distribution (Fig. 1).

3 Results

The uncertainty of an inventory can be expressed most conveniently as a percentage of total emissions. Following the guidelines of the Intergovernmental Panel on Climate Change (IPCC; Penman et al.

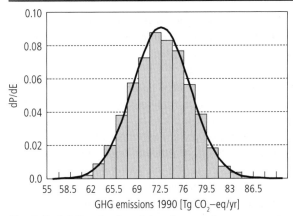

Fig. 1 Probability density of Austrian greenhouse gas emissions (data from Winiwarter and Rypdal (2001) in *column-shaped "bins"*) resembles the shape of a normal distribution function (*solid line*)

2000), we apply two standard deviations of the Monte Carlo output as estimate of the 95% confidence interval. The overall uncertainty of the Austrian inventory described by Winiwarter and Rypdal (2001) is just over 10% of the total emissions level. Uncertainty of the trend – the difference between 2 years – is expressed as percentage points (relating to the emissions level rather than the trend itself) and is thus not very sensitive to the trend or to the length of the period investigated. In the Austrian case specifically, the trend uncertainty determined for the period 1990–1997 was close to five percentage points, calculated as two standard deviations of the trend variability per level of emissions. This figure is clearly lower than the uncertainty of the emissions level, as emission sources contributing strongly to uncertainty in the level of emissions, but having identical emission factors in different years, will not contribute to the trend uncertainty (high covariance). This is important to note, as national obligations according to the Kyoto Protocol are expressed in terms of trends. While these obligations do not address uncertainty, a retrospective assessment will also obviously evaluate the confidence associated with their attainment/nonattainment. For such an analysis, a reliable estimate of trend uncertainties is possible even now, before 2010 emission inventories become available. As long as the principal emission pattern does not change, it will not make much difference which specific years are selected. While the trend may be quite different, the trend uncertainty for

Austria (and similar countries: see Monni et al. 2004; Rypdal and Winiwarter 2001) will remain close to five percentage points.

Sensitivity analysis allows the strongest contributors to total uncertainty to be identified. At the level of aggregation chosen for the uncertainty assessment, the largest contribution typically derives from an incomplete understanding of soils (Rypdal and Winiwarter 2001). Soil N_2O emissions are caused by microbial soil processes converting nitrogen compounds available in the soil. While the driving activity parameters like agricultural area or fertilizer input are known from statistical data, very different estimates have been made of the emission factor, the fraction of nitrogen added to soil that is actually emitted. When different authors derive their own conclusions from these estimates, even the resulting overall uncertainties are strongly affected (see below). Other important contributions to the overall uncertainty (here, specifically for Austria: Winiwarter and Rypdal 2001) are the uncertainty about the amount of solid waste (organic material that decomposes to produce methane) that is deposited and about the extent of land use change. Monni et al. (2004) consider peatland carbon dioxide (CO_2) for Finland to contribute even more strongly to the overall uncertainty than soil N_2O emissions.

4 Discussion

Reducing uncertainty is a key aim in terms of improving the applicability of inventories. An understanding of the mathematical relations involved allows several options to be identified that might facilitate the task of reducing uncertainty. We will now present and discuss the consequences of these.

4.1 Excluding Sources with High Uncertainty

Probably the simplest option for reducing the uncertainty of a national emission inventory is a redefinition of its system boundaries such that sources that make a high contribution to overall uncertainty need not be considered. As discussed above, emission inventories need not cover all sources within a country. Although definitions are to some extent arbitrary, they nevertheless result from political agreement and are identical for all countries.

Including a redefinition of sources in an inventory in order to lower overall uncertainty is technically feasible (even if politically difficult). This could relate not only to N_2O emissions from soils but also to sources like landfills and land use change. However, there are important considerations that discourage such a modification of the national inventory. First, it could exempt certain activities from a national reduction requirement, even though they are clearly caused by human actions. Furthermore, one main reason for fully covering all greenhouse gases and all sources is that abatement options may be cheaper for other gases than for carbon dioxide. In fact, it has been shown that the large contributors to uncertainty are among the most promising for cost-effective emission reductions (Klaassen et al. 2004).

4.2 Covariance and the Definition of Adequate System Boundaries

One particularly interesting aspect of error propagation is the effect of covariance on the additive terms of an emission calculation. While difficult to treat in a classical error propagation calculation, covariance can easily be dealt with by Monte Carlo methods. It can be shown that the relative error (twice the standard deviation divided by the mean value of a sample) assigned to two or more independent emission sources (i.e., without covariance) is higher than that of the total. In other words, adding independent information on emission sources to a total will decrease the uncertainty of the new total. As a result, it makes sense to break down a source category into small individual sources treated separately, as the overall uncertainty will decrease. This is different when the sources are correlated (i.e., a random shift in one of them will also affect the result of the other). One example of this is when an identical set of emission factors is applied to similar sources. If the emission factors tend to be inappropriately high/low, they will be too high/low for all sources. Adding differentiation to these dependent sources will not reduce the overall relative uncertainty, as they share an identical emission factor.

There may be different degrees of correlation or covariance; thus, the reduction of uncertainty may be affected in different ways. In the case of moderate (i.e., partial) covariance, overall relative uncertainty may be somewhat reduced, but not as much as in the case of independent sources.

It is also important to consider inverse correlation, which occurs when the uncertainty associated with a total is smaller than with one component of this total. For example, the national total fuel consumption statistics may be considered less uncertain than the statistics for a specific economic sector. In this case, the "rest" (i.e., all other fuel use in the country) can be calculated from the difference between this specific sector (albeit with high uncertainty) and the total (with much lower uncertainty). While the uncertainty of the "rest" will be high, it will take the opposite direction (inverse correlation) from the uncertainty of the specific sector, which is also determined by the low uncertainty of the total. Combining the individual sectoral uncertainties to an overall uncertainty necessarily yields the same result as that of the total–inverse correlation will cause relative uncertainty to decrease even more strongly than independent parameters would.

The effect of inverse correlation plays an important role in the land use and land use change sector. Full carbon accounting may involve components that are strongly inversely correlated. An attempt to assess uncertainty as if components were independent may lead to a significant overestimation of overall uncertainty (e.g., Nilsson et al. 2000). Moreover, the way boundaries are set between anthropogenic and natural sources can affect this uncertainty. A more comprehensive inclusion of sources might significantly decrease relative uncertainty, when overall information (e.g., on biomass growth within a certain biome) is available with higher accuracy. This topic has been discussed in more detail by Nilsson et al. (2007).

In a comparable way, the selection of system boundaries in emissions trading also affects the uncertainty involved. Monni et al. (2007) show that the choice of sources included in a trading system significantly influences uncertainties. A consistent trading scheme reflecting the individual uncertainties associated with respective sources has been devised by Nahorski et al. (2007), in whose approach system boundaries are mathematically transformed rather than shifted.

4.3 Significance of Subjective Interpretation of Uncertainty

When comparing studies of uncertainties in GHG emissions from different countries (Rypdal and Winiwarter 2001), the expert judgment of just one factor appears to be responsible for the much higher overall uncertainty of 20% presented by some developed countries (Norway, United Kingdom) compared to the roughly 10% for Austria (and also The Netherlands or the United States). It is the uncertainty of the emission factor on N_2O from soils, or rather the interpretation of this uncertainty, that drives the discrepancy, even if all approaches are supported by the literature. Since this comparison was published, new results on national emission inventory uncertainty have become available. However, these will not provide new elements as long as no new arguments are being advanced.

Soil N_2O, though extremely important, seems to be an exception. The available data indicate that, except for this source, inventories of industrialized countries show largely similar features, and the results presented for Austria can be considered representative. Sensitivity analysis shows that the resulting uncertainty depends mainly on a few input parameters. In almost all cases, the most sensitive contribution is the uncertainty in N_2O emissions from soils.

As N_2O soil emissions contribute consistently and strongly to uncertainty and, moreover, are responsible for the largest differences among national inventories, we will analyze this source sector in more detail. An investigation of this kind primarily aims to improve knowledge of the emission process and so reduce its uncertainty. However, an increased understanding of the magnitude and nature of the uncertainty will also assist in reducing differences between countries. Some of the more advanced efforts to improve knowledge on N_2O emissions from soils have been compiled by Leip et al. (2005).

An analysis of European N_2O emissions (Winiwarter 2005) reveals further differences in exactly how the official IPCC guidelines are applied (Houghton et al. 1997). According to the guidelines, there should be accounting of the different pathways of nitrogen input to soils. There is at least one country that reports the important pathway of applying animal manure as animal emissions rather than soil emissions. Although this does not affect the overall emissions balance, such a difference may influence the interpretation of country intercomparisons and the relative weighing of abatement measures. Again, this difference results from a subjective interpretation of the guidelines.

National data have been collected and used to calculate the N_2O from soil emissions at the level of the 15 "old" member countries of the European Union (Boeckx and van Cleemput 2001) based on default emission factors from the IPCC guidelines. These results are different from those in the national emissions reports, as they are not based on details available to the national inventory agencies. The differences, however, are not greater than the differences in independent approaches, like a regression model based on field measurements or process-oriented models (see Winiwarter 2005, for a more detailed discussion). Most remarkably, all these individual results are in a range of less than a factor of two for each individual country, while the uncertainty of this source has been estimated at two orders of magnitude (Houghton et al. 1997). Certainly, all the individual approaches influence each other, as the results are well known in the scientific community and any discrepancy needs to be well explained before being publishable. Implicitly, the large uncertainty margin may contain an element of (subjective) safeguarding to account for unknown systematic errors.

Further detailed analysis of this source sector and especially an elucidation of the processes leading to the emissions will help to further decrease inventory uncertainty or at least motivate inventory compilers to use the same assumptions for assessing uncertainty. Still, one must not expect uncertainty to completely disappear. While the uncertainty of an emission factor (perfectly describing the processes) is considered equally applicable to different years and does not affect the trend analysis (full correlation), the uncertainty of the activity number is considered independent. As there will be a high improbability of obtaining sufficient activity data from the past to improve uncertainty for the base year of an inventory, an important part of trend uncertainty will remain. It has thus been shown that trend uncertainty for the full inventory cannot decrease below three percentage points (Winiwarter and Rypdal 2001). Appropriate selection of the sources included in the emission inventories (specifically, exclusion of land use change sources) could, however, also reduce this uncertainty.

4.4 Rigid Accounting as a Method of Assessing Emissions

With trend uncertainties of several percentage points being typical of industrialized countries, a statistical reevaluation of the achievements of reduction targets of 6–8%, as formulated in the Kyoto Protocol, will provide mostly ambiguous results. While such a reevaluation is not part of an international agreement, considerations on post-Kyoto targets are likely to include an evaluation of this kind. Assuming that the trend uncertainties will also remain at the currently estimated level for the year 2010, and assuming that the trends will be in the predicted range, many countries will not be able to provide statistically significant conclusions on the achievement of their targets, and only a few will either clearly meet the target or fail to do so. Uncertainty assessment will often not allow a differentiation as to whether targets, in terms of real fluxes to the atmosphere, have been met.

Emission reductions as proposed by the signatories of the Kyoto Protocol are far too small to change the increasing trend of atmospheric GHG concentrations. Instead, the Protocol may only provide a first step in emission reduction, with further target setting to come. Consequently, it is neither the real atmospheric concentrations nor the real fluxes to the atmosphere that are the target of reduction. Instead, it is a fair share of burden distributed to individual countries. To safeguard a fair share, it is not necessary for the individual trend uncertainties for countries to be small. It is only necessary for any two countries to be treated the same way (i.e., for the uncertainty in the difference between their respective emission trends to be small).

Using the identical emission factors for two different years may not always provide the best available figures for emission assessment – or, in other words, it may not always bring the uncertainty in the emissions level to the lowest level possible. Nevertheless, the procedure ensures that the correlation of emission factors can be fully considered for trend uncertainty calculation, and this will remove an important part of the uncertainty, allowing trend uncertainty to be smaller than level uncertainty.

In a very similar way, applying the identical procedures for emission assessment in two different countries will allow methodological correlation (in emission factor assessment as much as in derivation of activity numbers) to take place. This possibly means a disadvantage in terms of the assessment of the level of emissions, as special national conditions cannot be fully accounted for. Still, the rigorous harmonization of input information for emission calculation will emphasize the statistical correlation of the inventories of any two countries. As a consequence, the uncertainty involved in the difference between their emission trends will become even smaller than the uncertainty of the emission trends themselves – as is required for fair burden sharing.

Decisions by country experts to select the ideal approach to assessing a country's emissions will provide very valuable information to this country's inventory, especially if specific national information is brought in that cannot be covered by generic guidelines. But, as discussed above, even a well-established inventory may not be able to assess the extent to which reduction targets have been met. It thus seems useful to focus instead on a different aspect, minimizing differences among countries so as to arrive at a low uncertainty of the differences between their emission trends. This can be accomplished by not applying individual approaches for countries and by harmonizing any subjective decisions that need to be taken.

We thus propose a rigid emission accounting system instead of a scientifically perfected emission inventory. Within this system, adherence to accounting rules rather than attempting to reflect a real material flow situation must be the first priority. The intercountry comparison ("fair burden sharing") – not the assessment of actual emissions – must be the ultimate goal. Such an accounting system needs to be based on scientific knowledge and could derive from existing information (Houghton et al. 1997), but the choice of parameters or approaches by country experts should be kept to the minimum. Once fixed, the system should be kept constant for the commitment period.

Removing this other large contributor to uncertainty, the subjective decision, will ensure more equal treatment of countries, even if uncertainty in terms of atmospheric fluxes remains high. As countries' statistical systems may have been established in a similar fashion, even the above-mentioned minimum trend uncertainty of three percentage points (introduced by the impossibility of fully readdressing historic data) may be reduced when assessing the uncertainty of trend differences between countries. The extended use of correlated input data will further remove uncertainties when considering trend differences, which are in fact differences between differences.

A rigid, scientifically based scheme has already been developed for other aspects of GHGs. The greenhouse warming potential (GWP) is defined as the mass of CO_2 emissions that, over a 100-year period, would contribute the same radiative effect as one mass unit of the compound in question. This factor is commonly used, even if its exact magnitude for the respective GHGs is still a matter for discussion, and it may be subject to future change.

Developing and establishing such a rigid emission accounting system certainly requires considerable resources. Formerly, resources available for emission inventories were scarce with respect to the financial stakes involved in emissions trading schemes. Emissions estimates were derived from data collections (statistics) that were established for completely independent reasons and could be considered unbiased for the purposes of estimating emissions. As such emissions estimates will now become tradable assets, there is even more reason to convert the system to an accounting system. Evaluation, control, and improvements in emission inventories will require efforts that need to be seen in perspective (and in proportion to the assets covered). A reasonable emission accounting system will provide confidence in the emission inventory and even more protection for such assets.

5 Conclusions

While a few options exist for reducing uncertainty in national greenhouse gas emission inventories, there is little possibility of decreasing uncertainties to a level that is much lower than current emission reduction targets. At least among industrialized countries, the inventory uncertainties that have been presented to date are not a measure of inventory quality but rather the result of subjective decisions on the part of the respective country experts. Any attempt to use the uncertainty as a tool for justifying adjustments to an inventory is thus void – see Gillenwater et al. (2007) for a thorough review. More understanding regarding the uncertainty of national inventories can be brought in by new – possibly independent – studies, not by applying existing uncertainty denominators to yet another country. As long as no new input information regarding the uncertainties of the respective inventory input is provided, the assessment of country inventory uncertainties for additional countries will, for the most part, provide results that are similar to the uncertainty assessments that are currently available; or it will yet again show the contribution of subjective assumptions.

Instead of relying on uncertainty as a subjective ingredient in emission inventories, we suggest reducing uncertainty to the minimum by reducing the individual choice possible in the compilation of national GHG emission inventories. While this reduction in choice may not lead to a fully adequate treatment of emissions (as rigidity will not allow for a specific source representation), any lack of adequacy will provide a common basis for different countries. An intercountry comparison will then yield comparable results at a low uncertainty level and may allow for fair burden sharing beyond current emission–reduction obligations.

The quantification of uncertainty is needed at a different stage of the overall procedure of setting and checking emission targets. A periodic review of the emission reduction targets, following the need to limit anthropogenic climate forcing, will have to consider inventory uncertainty. This will allow an evaluation of whether objectives have been met as well as facilitating validation with external information like atmospheric measurements. The evaluation of uncertainty will be even more necessary if emission-calculating algorithms are accepted that are less than perfect as a result of the rigorous accounting technique proposed here. As part of such a review, designed to set new targets on emission reduction, the rules for emission accounting could also be adapted to reflect the latest state of science.

References

Boeckx, P., & van Cleemput, O. (2001). Estimates of N_2O and CH_4 fluxes from agricultural lands in various regions in Europe. *Nutrient Cycling in Agroecosystems, 60*, 35–47.

EEA (2004). *Joint EMEP/CORINAIR atmospheric emission inventory guidebook (3rd edition, September 2004 update)*. Technical Report no. 30, Copenhagen, Denmark: European Environment Agency. See http://reports.eea.eu.int/EMEPCORINAIR4/en.

Gillenwater, M., Sussman, F., & Cohen, J. (2007). *Practical applications of uncertainty analysis for national greenhouse gas inventories* (in press).

Houghton, J. T., Meira Filho, L. G., Lim, B., Treanton, K., Mamaty, I., Bonduki, Y. et al., (Eds.) (1997). *Revised 1996 IPCC guidelines for national greenhouse gas inventories*. Bracknell, UK: IPCC/OECD/IEA, Meteorological Office.

Klaassen, G., Amann, M., Höglund, L., Wagner, F., & Winiwarter, W. (2004). *Medium-term multi-gas mitigation strategies*. Report to IIASA side event at the Conference of Parties (COP10), held on 8 December 2004 in Buenos Aires, Argentina.

Leip, A., Seufert, G., & Raes, F. (Eds.) (2005). *N₂O emissions from agriculture*. Report EUR 21675 EN of the European Communities, Luxembourg.

Monni, S., Syri, S., Pipatti, R., & Savolainen, I. (2007). *Comparison of uncertainties in different emissions trading schemes* (in press).

Monni, S., Syri, S., & Savolainen, I. (2004). Uncertainties in the Finnish greenhouse gas emission inventory. *Environmental Science and Policy, 7*, 87–98.

Nahorski, Z., Horabik, J., & Jonas, M. (2007). *Compliance and emissions trading under Kyoto Protocol: Rules for highly uncertain inventories* (in press).

Nilsson, S., Shvidenko, A., & Jonas, M. (2007). *Uncertainties of the regional terrestrial biota full carbon account: A systems analysis* (in press).

Nilsson, S., Shvidenko, A., Stolbovoi, V., Gluck, M., Jonas, M., & Obersteiner, M. (2000). *Full carbon account for Russia*. IIASA Interim Report IR-00-021, Laxenburg, Austria: International Institute for Applied Systems Analysis.

Penman, J., Kruger, D., Galbally, I., Hiraishi, T., Nyenzi, B., Emmanuel, S. et al., (Eds.) (2000). *Good practice guidance and uncertainty management in national greenhouse gas inventories*. Report of the IPCC National Greenhouse Gas Inventories Programme, Institute for Global Environmental Strategies, Kanagawa, Japan.

Rypdal, K., & Winiwarter, W. (2001). Uncertainties in greenhouse gas inventories – evaluation, comparability and implications. *Environmental Science and Policy, 4*, 107–116.

UNFCCC (2004). *Guidelines for the preparation of national communications by parties included in ANNEX I to the convention, Part I: UNFCCC Reporting Guidelines on Annual Inventories (following incorporation of the provisions of decision 13/CP.9)*. Report of the Twenty-first Session of the Subsidiary Body for Scientific and Technological Advice, held in Buenos Aires, Argentina. See http://unfccc.int/national_reports/annex_i_ghg_inventories/reporting_ requirements/items/2759.php (last accessed July 2006).

Webber, P. H., & Fleming, P. D. (2002). Emission models and inventories for local greenhouse gas emissions assessment: Experience in the East Midlands Region of the UK. *Water, Air & Soil Pollution. Focus, 2*, 115–126.

Winiwarter, W. (2005). *The extension of the RAINS model to greenhouse gases: N₂O*. IIASA Interim Report IR-05-055, Laxenburg, Austria: International Institute for Applied Systems Analysis.

Winiwarter, W., & Rypdal, K. (2001). Assessing the uncertainty associated with national greenhouse gas emission inventories: A case study for Austria. *Atmospheric Environment, 35*, 5425–5440.

To develop an adjustment scheme, we must, however, develop a more specific definition of environmental integrity. It is reasonable to begin the discussion with the targets set by the Kyoto Protocol (Annex B) for each participating developed country for the first commitment period. Suppose we start by defining as our goal that we want to be confident that, when countries report emissions inventories that nominally are in agreement with their commitments under the Protocol, the countries truly are, if not in compliance, at least within a given tolerance of complying with their commitments. Thus, we might consider an adjustment based on uncertainty as described in Definition 1.

Definition 1 Compliance with Emission Targets: Attain a reasonable level of confidence that countries have actually achieved the target emissions levels stated in their commitments under the Kyoto Protocol and are in compliance.

To implement this definition, we ask three questions: (1) Would we consider it acceptable if actual emissions *exceeded* the target emissions commitment by some fractional or percentage amount? (2) How much is that amount? (3) How confident do we want to be in our result? If we assume that we know the magnitude of uncertainty surrounding the inventory estimate (an assumption we revisit later in the paper), this definition suggests that emissions inventory estimates would be adjusted upward to take into account the uncertainty of the estimate. In particular, the assumption would be that we want to ensure that, given a reasonable level of confidence, actual emissions do not exceed estimated emissions by more than a specified amount.[8]

Table 1 illustrates the types of adjustment that this definition might imply, based on various quantified levels of uncertainty in an inventory estimate, on the amount of confidence we want to have in our results,

and on the percentage amount by which actual emissions could exceed the emissions commitment (i.e., the target level of emissions) before we were uncomfortable with the result.[9] For example, if emissions estimates are 50% uncertain and we want to be 90% certain we have not exceeded our emission target by more than 10%, we would adjust the emissions inventory estimate upward by 20%. The adjusted emissions value would then be compared against the target emissions value to determine compliance. This adjustment provides a margin of safety; that is, a country would effectively need to reduce emissions by more than its commitment in the Kyoto Protocol to remain in compliance with commitments. The higher the level of uncertainty surrounding the emissions inventory estimate, the greater the adjustment that would be required. Similarly, the greater the degree of confidence we require, the greater the adjustment.

The analyses in Table 1 and later in this paper make the simple assumption that the uncertainty distributions are normal and symmetric about the inventory estimate (i.e., there is no bias). In theory, a normal distribution cannot be exactly correct, because negative emissions values are impossible, but this error will be negligible if the probability of a negative value is sufficiently small. For GHG emissions inventories, normal, log-normal, uniform, triangular, and beta distributions have been used to model uncertainty distributions, often truncated to force the values to be within a plausible range. While we could carry out exactly the same analyses for other choices of uncertainty distributions, the normal distribution is a sufficient choice to illustrate our conclusions. In principle, by using a Monte Carlo simulation, all of the numerical approaches described in this paper could be applied to any given uncertainty distributions for a national GHG inventory.

---

[8] Throughout this discussion, we assume that probability distributions for estimated emissions or emission reductions are normal and that the shape of the probability distribution of emissions for each country or source does not change significantly as emissions are reduced. This entire analysis also ignores the possibility that we might *underestimate* actual emission reductions (i.e., this analysis assumes that the purpose of investigating uncertainty is to ensure that we do not *overestimate* actual emission reductions).

[9] Given the uncertainty ($u\%$) range (assumed to be the end points of a 95% confidence interval) around estimated emissions (E), and assuming a normal distribution, the standard deviation of the distribution equals approximately $u\% E/ (1.96)$. If we are willing to accept that our emissions could be up to $p\%$ higher than the nominal emissions commitment, then the probability that the actual value lies *below* an upper bound of $(100+p)\% E$ can be calculated from the table for a normal error integral found in standard statistics textbooks or using standard statistical software (including Excel). See, for example, Appendix A in Taylor (1997).

Table 1 Ratio of adjusted emissions to estimated emissions

Confidence[a]	Uncertainty of emissions inventory		
	20%	50%	80%
95%	1.06	1.30	1.52
90%	1.03	1.20	1.39
85%	1.01	1.15	1.30
80%	n/a	1.10	1.22

[a] Confidence that actual emissions will not exceed emissions estimate by more than 10%.

Sources: Sussman, 1998, and Sussman, Cohen, and Jayaraman, 1998

The definition of environmental integrity proposed above focuses on only one aspect of emissions uncertainty: the uncertainty of current-year emissions estimates as they are reported for compliance purposes. However, the emissions estimate for the base year – from which the commitment level for a country is calculated under the Kyoto Protocol – is subject to uncertainty that is likely to be of similar or greater magnitude than the uncertainty of the emissions estimate for a commitment period.[10] The uncertainty in a country's base-year emissions or removal[11] estimates may be greater than that during the commitment period because countries will, hopefully, have made improvements in their inventory over time, some of which cannot be fully implemented by recalculating the base – year estimate.

We can broaden the definition of environmental integrity to take into account the influence of uncertainty in both the base year and the current inventory year. Focusing on emission reductions, rather than on emissions, is one way of accomplishing this. In particular, we can argue that it is more important to ask whether or not we have reduced emissions (and, in the case of the Kyoto Protocol, achieved the emission reductions to which countries

have committed) than to ask whether emissions are actually what we think they are. Moreover, as the uncertainty surrounding the level of emissions is not identical to the uncertainty surrounding the absolute (or relative) level of emission reductions, we can develop a second definition.

Suppose that a country has agreed to reduce emissions to a target level in a given year (or set of years). If estimated emissions in that time period equal the target level, how confident can we be that emissions have actually been reduced by an amount equal to the difference between base-year emissions and estimated emissions in the target period? Put another way, how confident can we be that estimated emission reductions are not smaller than we think they are or, at least, that they are not "off" by more than a certain amount. Following this line of reasoning, we might choose to define environmental integrity along the lines of Definition 2.

Definition 2 Achieving Emission Reductions: Achieve a reasonable level of confidence that countries have actually achieved the emission reductions, measured relative to base-year emissions, stated in their commitments under the Kyoto Protocol and are in compliance.

To implement this definition, we need to ask, (1) Would we consider it acceptable if actual emission reductions were to fall *below* the committed level of reductions by some fractional or percentage amount? (2) How much is that amount? (3) How confident do we want to be in our result? If we assume that we know the magnitude of uncertainty surrounding the estimated emission *reductions*, this definition suggests that estimated emission reductions would be adjusted downward to take into account the uncertainty of the estimate. However, the result can be compared more easily with the results in Table 1 if we ask how the *emissions estimate* for the commitment period would have to be adjusted upward to ensure that, given a reasonable level of confidence, actual emission reductions do not fall below estimated reductions by more than a specified amount (which could be zero). Again, the conclusion is that emissions estimates would be more heavily increased for more uncertain inventories.

We can construct Table 2 in a manner analogous to Table 1, but this time beginning by looking at

---

[10] Uncertainty may also differ (and in fact may be lower) in the base year because of policy and political changes over time, including the effects of economic reforms. These changes can affect the definition of what types of sources and sinks are included in the emissions estimate.

[11] A reviewer pointed out that removals are not normally accounted for in the base year under the Kyoto Protocol, except for some 3.4 activities.

uncertainty in emission reductions.[12] Our goal is to provide a level of confidence that our emission reductions have actually been achieved. Given that goal, we can ask what adjustment should be made to the nominal emissions inventory for the commitment period in order to compensate for the uncertainty of emission *reductions*. Suppose that emissions in a commitment year must be 7% below emissions in the base year for compliance (a number that translates into a target absolute quantity of emission reductions). Then, if quantified emission reductions are 50% uncertain and we want to be 90% confident that we have achieved at least 90% of the target quantity of emission reductions, the emissions inventory estimate should be adjusted upward by 3%. The adjusted emissions estimate is then compared with the target level to determine compliance.[13]

The two approaches have some similarities. Both approaches focus on ensuring a reasonable level of confidence with which we achieve externally defined goals; that is, quantified emissions or emission reductions for a target year or period, such as the first commitment period under the Kyoto Protocol. By adjusting emissions estimates to account for uncertainty, both approaches provide a concrete incentive for countries to reduce estimated emissions below nominal emission requirements. Thus, both approaches increase the confidence that we can have in our global emissions estimates, by adjusting the estimated emissions to account for uncertainty. They also provide an incentive for countries to reduce the uncertainty of their emissions estimates over time, in order to reduce the magnitude of the adjustment and so move estimated emissions closer to the nominal commitment level.

Which approach is more stringent? Assuming for the moment that the uncertainty of the emissions estimate

Table 2 Ratio of adjusted emissions to estimated emissions

Confidence[b]	Uncertainty of emission reductions[a]		
	20%	50%	80%
95%	1.01	1.04	1.15
90%	1.00	1.03	1.08
85%	1.00	1.02	1.04

[a] Emission reductions for compliance assumed to be 7% below baseline level.

[b] Confidence that actual emission reductions equal at least 90% of estimated reductions.

Source: Sussman et al., 1998

(in Table 1) is the same as the uncertainty of the emission reductions (in Table 2), then the fractional adjustments are much larger under Definition 1 (in Table 1) than under Definition 2 (Table 2), because in the former case the definition focuses on the absolute level of emissions, which is a much larger number than the absolute reduction in emissions (the focus of the latter definition). Whether or not this is a legitimate assumption, however, and the relationship between uncertainties in emissions relative to uncertainties in emission reductions, are not addressed here.

These are only two of many different possible environmental goals and associated statistical adjustments that could be performed. Other environmental goals could be employed that would result in larger (or smaller) adjustment factors. For example, if our environmental goal were to have confidence in the environmental impact of meeting the target commitments, we would want to apply the adjustment factor to the base-year estimate on which the target commitments are based, and so indirectly adjust the actual target commitment (downward, in this case, to reflect uncertainty). To the extent that inventory uncertainty is likely to decline over time as inventory methods improve, this type of adjustment may make sense. We might, therefore, also want to look at the uncertainty in trends and determine if adjustments to actual emissions estimates are also warranted, or if the adjustment to the base year is sufficient to meet our environmental goals.[14] Another alternative approach might be to

---

[12] It may not be immediately obvious how to calculate the uncertainty of emission reductions, as it will depend not only on uncertainty in the base and current year, but also on correlations between the two uncertainty estimates, since the factors that produce bias in one year may produce bias in another year. Winiwarter and Rypdal (2001) have looked at trend uncertainties for the Austrian inventory.

[13] Constructing Table 2 requires two steps: (1) making necessary assumptions (e.g., about the uncertainty of emission reductions and the required level of confidence) and calculating the necessary adjustment in emission reductions to provide that level of confidence, and (2) translating the adjustment to emission reductions into an adjustment to the emissions estimate.

[14] The UNFCCC approach uses adjustments to both the base and current year. Again, these adjustments are primarily designed to encourage the use of good practice inventory methods (while also providing some environmental benefit) and are not related to the uncertainty of the overall inventory or of a specific source category for a particular country.

focus on the commitment level rather than the inventory estimate (i.e., how country emission targets would need to be adjusted *downward* in order to ensure that we are confident that we are meeting the current emissions limits specified in the Kyoto Protocol).

2.2 Characteristics of the Adjustment Factor and Implications for the Uncertainty Analysis

The approaches described above result in potentially large adjustments to the emissions inventory. Given the political debate that raged over the targets for commitments under the Kyoto Protocol – with the average across all countries for the first commitment period finally settling at around 5% below base-year emissions-additional reductions of even 1% could have serious political ramifications.[15] An adjustment factor could also have significant associated control cost implications for countries that face additional reductions. Thus, for an adjustment factor to warrant the additional economic costs, we suggest that the factor should possess the following characteristics (many of which are the same characteristics that the national inventory should possess):

- It should meet clear environmental goals and be statistically justifiable given those goals (as described in Section 2.1).
- It should be applied fairly and objectively across countries and source categories (i.e., the method for calculating the factor should rely on data that can be reviewed and verified).
- It should be comparable across countries (i.e., not be subject to inherent variability based on unexplained or unexamined differences in methodology, expert judgment, or expenditures).
- It should be administratively feasible and not burdensome, so that it is practical to calculate and apply the factor.

---

[15] In addition to general political considerations and the feasibility of negotiating an international system of adjustments that would require reductions beyond those already agreed to (as in the Kyoto Protocol), such a system could raise equity concerns if poorer nations were also those with greater uncertainty, especially if this were primarily due to the source composition of their inventory. In particular, nations with inventories that have a large component of non-energy sources will tend to have greater uncertainties that would be relatively expensive to reduce.

- It should not be easily manipulated by countries acting in their own self-interest.
- It should not influence market values in a way that (unintentionally) impedes allowance trades between countries.

In large part, the answer to whether or not the adjustment factor can meet these criteria will depend on the characteristics of the uncertainty analysis. In the context of adjustment factors, the uncertainty estimates for the GHG emissions inventory will face several challenges.

Box 1: Sources of Uncertainty in GHG Inventories
Uncertainties associated with GHG inventories can be broadly categorized into *scientific* uncertainty and *estimation* uncertainty. Scientific uncertainty arises when the science of the actual emission and/or removal process is not completely understood. For example, the process of indirect N_2O emissions associated with nitrogen-containing compounds that are first emitted to the atmosphere and then deposited on soils involves significant scientific uncertainty.

Estimation uncertainty arises any time GHG emissions are quantified. Therefore, all emission or removal estimates are associated with estimation uncertainty. Estimation uncertainty can be further classified into two types: *model* uncertainty and *parameter* uncertainty. Model uncertainty refers to the uncertainty associated with the mathematical equations (i.e., models) used to characterize the relationships between various parameters and emission processes. For example, model uncertainty may arise either from the use of an incorrect mathematical model or from the use of an inappropriate input in the model.

Parameter uncertainty refers to the uncertainty associated with quantifying the parameters used as inputs (e.g., activity data and emission factors) to estimation models. Parameter uncertainties can be evaluated through statistical analysis, determinations of the precision of measurement equipment, and expert judgment. Quantifying parameter uncertainties and then estimating source category uncertainties based on these parameter uncertainties is typically the primary focus of most national inventory agencies.

Box 2: Inventory Uncertainty versus Regulatory and Market Uncertainties
Inventory uncertainty relates to the uncertainties in the quantified emissions (or removals) reported in GHG inventories (see Box 1). In contrast, regulatory uncertainties relate to uncertainties in how current or future regulatory rules will affect compliance determinations, and market uncertainties relate to the uncertainties in future allowance prices, mitigation costs, and transaction costs. Both regulatory and market uncertainties are largely independent of inventory uncertainties.

For example, the rules for an emissions trading scheme specify the methodologies that are acceptable for estimating emissions. Emission allowances of a quantity equivalent to these emissions must then be surrendered for compliance purposes. Markets only respond to uncertainty in the value of the traded item-whether it is what it says it is. Thus, an allowance will be worth the price of one ton of emissions, if the rules of the trading scheme say it is worth 1 t of emissions. There will be no regulatory uncertainty about its value on the market if the rules are clear, regardless of the uncertainty in the emissions inventory estimate. If, however, there is uncertainty about the rules that define the quantity of emissions for which an allowance must be surrendered, then regulatory uncertainty will affect the value of an allowance in the market.

Regulatory and market uncertainties have enormous impacts on emissions trading markets. However, these markets are relatively ambivalent about inventory uncertainties, unless they are perceived to have an impact on emissions trading rules (e.g., they are the basis of emissions trading ratios). Policy makers and the public, however, may show concern about inventory uncertainties if they perceive them to be high enough to cause the compliance process to lack credibility or environmental efficacy. Thus the rules of an emissions trading scheme are the conduit though which inventory uncertainties can affect regulatory and market uncertainty.

The subjectivity of uncertainty estimates First, for some source categories, the uncertainty estimates possess a strong subjective component. The inventory is subject to several types of uncertainty (see Box 1 for a discussion of inventory uncertainty and Box 2 for the distinction between inventory uncertainty and regulatory and market uncertainties). Of these, scientific uncertainty and model uncertainty are particularly difficult to quantify, because they must rely heavily on expert judgment regarding inherent uncertainties and potential biases in the estimation methodology. Moreover, expert judgment will be a significant and unavoidable component of uncertainty estimates for national inventories, since the measurements and sample data needed to produce probability distributions will rarely exist for the emission factors and activity data used to produce GHG emissions inventories.

While for some scientific exercises it is possible to collect rigorous statistical data that can be used to estimate the statistical uncertainty of a parameter,[16] it

is often impossible to collect similar sample data for many of the national statistics used in inventories. Often only a single data point will be available for a parameter (e.g., tons of coal purchased). It is not meaningful to repeatedly collect independent sets of national statistics for the same year. Instead, we are often given a single emission value or activity factor that supposedly is a census of the entire population rather than a statistical sample, and so is unrepeatable. Our uncertainty estimate in this case represents an assessment by one or more experts of the probabilities that the estimate differs from the true value by "x," partly based on the experts' general experiences of similar estimation problems and inventory data and partly based on the experts' understanding about the country-specific inventory, such as possible double- or undercounting of emissions.

The subjectivity of the estimates for some source categories (and, hence, for the inventory overall) has several important consequences. Because of the subjectivity of inputs to the uncertainty estimation process and the reliance on expert judgment, it will be difficult and time-consuming to prepare a detailed uncertainty estimate that is totally transparent and reproducible, and that thoroughly documents all the expert judgment necessary to produce a comprehensive analysis. The analysis, therefore, will not be easily verified by the international community. The difficulty of verifying uncertainty estimates also raises the potential problem that countries may manipulate the uncertainty estimates for their inventories to their own advantage.

Moreover, because expert judgment will vary with the expert and according to his or her familiarity with the inventory data, it will vary from country to country, and even among source (or sink) categories within a country. Therefore, uncertainty estimates will not be comparable across countries, raising the question of whether the adjustment mechanism is an equitable one. Reliance on experts can produce considerable variability in the uncertainty estimates across countries using different experts. Rypdal and Winiwarter (2000) report that, for N_2O, uncertainty estimates vary dramatically – by two orders of magnitude – across existing country estimates. While differences in data and methods account for a portion of the difference, a large part of the difference is attributable to differences in the subjective assessments provided by expert judgment (Morgan & Henrion, 1990).

---

[16] Statistical uncertainty results from natural variations (e.g., random human errors in the measurement process and fluctuations in measurement equipment). Statistical uncertainty can be detected through repeated experiments or sampling of data.

Of particular concern are cases where the expert's uncertainty estimate is high. In these cases, the available information and data are likely to be extremely limited, and therefore an expert may not be able to quantify the uncertainty much beyond an assertion that the estimate of the parameter is unreliable. For example, an estimated uncertainty of 80% by expert judgment might mean the same as an estimate of 150% or more. (An estimate of 80% by one expert might be the same as one of 150% by another.) In practice, applications of uncertainty analyses should probably be limited to cases where the uncertainty is reasonably low (e.g., less than 80%), or where expert judgment plays a small role.

Systematic bias in uncertainty estimates Second, for some source and sink categories, systematic biases[17] may be the primary cause of uncertainty, especially for activity data (e.g., underreporting by companies or black market activities).[18] Therefore, countries will usually have to rely on expert judgment for the majority of their parameter uncertainty estimates.[19] Even with the most rigorous expert elicitation

protocol, it is difficult to obtain judgments in a comparable (i.e., unbiased) and consistent manner across parameters, source categories, countries, and inventory reporting years. Some experts will inherently tend to be optimistic about the quality of data, and others will tend to be pessimistic.[20] Thus, there may not only be a wide uncertainty band around the mean estimate of uncertainty, but the mean estimate itself may be inaccurate (i.e., subject to bias).

Availability of uncertainty estimates Finally, as most countries have not, thus far, undertaken detailed and rigorous uncertainty analyses, reliable estimates of inventory uncertainty are not generally available. An adjustment factor based on country-level uncertainty estimates would require considerable additional resource expenditures for each country that is party to the Kyoto Protocol, as well as considerable resources expended in verifying the estimates internationally. The additional expenditure would be much less, however, than has been expended in producing the inventory itself, at least once the basic structure of the analysis has been developed and implemented. Setting up these initial structures could, however, be time-consuming as well as resource intensive, which would certainly delay trading between countries and impede compliance activities (since countries do not know their actual inventories until they have calculated the adjustment factors). Such a situation could also increase the potential for disputes between an expert review team and a Party because of the subjective elements of the uncertainty analysis.

The Kyoto Protocol process The adjustment process that is under development under the Kyoto Protocol avoids some of these problems. While essentially punitive in nature (i.e., designed to encourage countries to follow "good practice"), it also acts to produce environmental benefits (i.e., the adjustment factor works to increase current-year emissions estimates or reduce base-year emissions). The Kyoto

---

[17] Systematic parameter uncertainty occurs if data are systematically biased. In other words, the average of the measured or estimated value is always less or greater than the true value. Biases arise, for example, because emission factors are constructed from non-representative samples, not all relevant source activities or categories have been identified, or incorrect or incomplete estimation methods or faulty measurement equipment have been used. Because the true value is unknown, such systematic biases cannot be detected through repeated experiments and, therefore, cannot be quantified through statistical analysis. However, it is possible to identify biases and, sometimes, quantify them through data quality investigations and expert judgments.

[18] There are cases where cause and direction of a specific systematic bias may be known for a national statistical dataset, but for reasons of resource and time limitations or political constraints they cannot be quantified or corrected for in the official national statistics. Therefore, arguing that known systematic biases can be corrected for ignores the real complexities of collecting national statistical data.

[19] The role of expert judgment can be twofold: First, expert judgment can be the source of the data that are necessary to estimate the parameter. Second, expert judgment can help (in combination with data quality investigations) to identify, explain, and quantify both statistical and systematic uncertainties. It is also important to recognize that it is difficult for experts to distinguish between statistical uncertainty and systematic biases. Therefore, elicited estimates of uncertainty tend to incorporate both.

---

[20] For example, in the United States an early estimate of the uncertainty in methane emissions from manure management based on expert judgment was ±15%. The following year, improvements were made to the methodology to account for more regional differences and corrections were made to some activity data. The resulting change in the overall emissions estimate was 60%.

Protocol process uses an approach similar to that of Definition 1 and Table 1 in this paper, but with different parameters. For emissions commitments, rather than applying the adjustment to all inventories, they apply the adjustment only in cases where an expert review panel finds enough problems with a country's inventory estimate to justify an adjustment to the inventory value. In such cases, after replacing a Party's estimate for an individual source or sink category with one from a review team (representing the central tendency, such as a mean or median), the review team makes a further, punitive adjustment to account for uncertainty. The adjustment uses the uncertainty estimates from the IPCC Guidelines, rather than country-specific uncertainty distributions.

Instead of the 10% leeway illustrated in this paper, the Kyoto Protocol approach allows 0% leeway – no leeway – so that the confidence level equals the probability of not exceeding the target. For our approach, with an illustrative 10% leeway, the confidence level equals the probability of not exceeding the target plus 10%. In contrast to illustrative 80–95% confidence levels used in this paper, the UNFCCC approach prescribes a 75% confidence level for upward adjustments of the commitment period emissions estimates (and 25% for downwardly adjusting base-year emissions estimates). Finally, the Kyoto Protocol approach assumes that emissions are log-normally distributed; the calculations in this paper assume normal distributions.

The Kyoto Protocol process avoids several of the problems of the country-specific adjustment factor described above. In particular, because adjustment factors are uniform across countries (if applied), the process avoids some of the issues of comparability, subjectivity, and gaming that could occur. The process also involves lower administrative costs, because fewer resources are expended on calculating country-specific uncertainty and the adjustment is only applied in select cases. The approach can be implemented more rapidly, so that countries will know more quickly what their inventories for a given year are. The environmental improvements of such a system are relatively low, however, since adjustments are only applied in specific, limited circumstances, and the process for deciding when adjustments are needed is itself extremely subjective and potentially political. Further, because the process is not designed with a clear and stated environmental goal in mind,

and because it does not use country-specific factors, it is unclear whether it is the most cost-effective means of obtaining environmental improvement.

Implications for adjustment factors in practice Where does this discussion leave us? Clearly, some system of adjustment factors would provide environmental improvement and increase our confidence that emission targets were being met. A country-specific set of adjustment factors that is applied across all countries would provide more confidence that targets were being met and would be statistically justifiable. Such a system would require a new and rigorous international system for reviewing and officially certifying uncertainty estimates. An adjustment factor applied to all countries, regardless of whether it is country specific or uniform, could also result in the largest environmental improvements. The choice between country-specific adjustment factors, which can be difficult to administer fairly, and uniform adjustment factors, which fail to reflect differences across national inventories, depends both on the environmental improvements each offers (i.e., how well the factors meet environmental goals or stated policy goals) and on the strengths and drawbacks of each approach.

3 Adjustments to Emissions Trading Ratios Based on the Uncertainty of Emissions

Now that the Kyoto Protocol has entered into force, developed (i.e., "Annex B") countries – excluding the United States and Australia, which have not ratified the Protocol – are legally committed to reduce their GHG emissions to specific negotiated target levels during the first 5-year commitment period (2008 through 2012).

In addition to meeting their commitments by reducing domestic emissions, Annex B countries can engage in three alternative market mechanisms that allow Parties to the Kyoto Protocol to purchase emission reductions from other Parties. The three mechanisms are (1) emissions trading, which permits buying and selling emission allowances among Annex B countries; (2) the Clean Development Mechanism (CDM), under which developed countries can undertake emission reduction projects in developing countries and use the emission reductions to offset their commitments; and (3) Joint Implementation (JI), which allows for project-related emission reductions within Annex B countries (e.g., mitigation projects in EIT[21] countries can produce emission reductions for developed nations). While the focus of this section is on international emissions trading, there are potential lessons for sales of project-related emission reductions under the CDM and JI.

In an emissions trading system, an administering authority generally sets quantified limits (referred to as rights, obligations, or permits) on the emissions of participants in the system. Participants can then transfer these rights, obligations, or permits from one participant to another (generally by buying and selling), subject to any restrictions set by the administering authority. Emissions trading systems are frequently referred to as "tradable allowance" systems, because participants must hold emission allowances, which give the owner the right to emit a specified physical unit of emissions, in sufficient quantity to cover actual emissions. Many consider emissions trading systems to be an attractive alternative to fixed emission limits because, in appropriate circumstances, they can reduce the overall cost of achieving an environmental goal. In a trading system, participants have flexibility in how they meet their obligations: they may choose either to take actions to reduce emissions or to purchase additional permits (if it is cheaper to do so). Thus, participants with lower costs of reducing emissions can undertake additional reductions and sell excess allowances to entities for whom the cost of reducing emissions is higher.

Under the Kyoto Protocol, Annex B countries can engage in emissions trading. The quantified limit for a country is its assigned amount (AA), and the instrument that is traded between countries is an assigned amount unit (AAU). As with other emissions trading programs, much of the debate in international GHG trading has centered on the impacts of trading on mitigation costs and the cost-effectiveness of emission reductions; on issues of the equitable division of responsibility for emission reductions; and on the design of domestic, facility-level emissions trading programs to support international trading.

Another issue in GHG emissions trading is the uncertainty of emissions data on which trades are based. Some analysts have suggested using trading ratios that are adjusted to reflect uncertainty, or even excluding highly uncertain emission sources from trading altogether, on the grounds of potential harm to the environment. Arguments have been made along these lines not only for the allowances traded between Parties (AAUs), but also for the instruments utilized by the project-based mechanisms of the Kyoto Protocol, the CDM (in which the tradable instrument is a certified emission reduction, [CER]) or JI (in which the tradable instrument is an emission reduction unit [ERU]).

The argument commonly made for prohibiting, or at the least placing a lower value on, emissions trades involving emissions from source categories for which the emissions estimates are highly uncertain is based primarily on the environmental harm that can be caused. If the uncertainty of the emissions estimate is high-or poorly understood-for some source or sink category, then emissions between certain and uncertain sources should not be traded (i.e., bought and sold) on an equal basis. For example, if society allows increased emissions from a source category with very low uncertainty in its emissions estimate to be offset by an equal quantity of emission reductions from a source category for which the emissions estimate is highly uncertain, we may not be sure that we have actually reduced emissions. Thus, the argument goes, any emission reductions or excess emission allowances from uncertain sources should be sold more cheaply (i.e., be worth less) than emission allowances or reductions from more certain sources. The trading ratio between allowances for certain and uncertain sources, therefore, is essentially less than 1: a given quantity of uncertain allowances will be equivalent to fewer certain allowances.

This approach of adjusting trading ratios to account for uncertainty has generally been adopted by watershed nutrient credit trading programs in the United States (King & Kuch, 2003). In these programs, nutrient discharges by diverse sources do not trade on an equal pound-for-pound basis. Rather, the trading ratio is based on the expected "risk-adjusted" outcome of trades; that is, on the certainty that a trade will actually result in decreased nutrient discharges and improved water quality.

---

[21] Economy in transition (EIT) is a term used under the UNFCCC to refer to the countries of the former Soviet Union and related East European satellite nations that are now undergoing a transition to a market-based economic system.

For example, the impact on nutrient levels, in the receiving water body, of changes in end-of-pipe nutrient discharges by a point source, such as a wastewater treatment facility, will be relatively certain. However, the impact on water quality of changes in land-management practices by non-point sources, such as farms, will be far less certain.[22] Thus, a wastewater treatment facility seeking to offset existing levels of nitrogen discharge may need to buy three or four pounds of non-point discharge reduction credits to offset each pound of nitrogen they are allowed to discharge, implying a trading ratio of 1:3 or 1:4. Note that, in this type of nutrient trading program, there is no accepted trading ratio for point and non-point trades. Each trade must be evaluated on its individual merits and approved by the regulatory authority, a process that can increase administrative costs (for both the traders and the administering authority) and the time required to finalize a trade (King & Kuch, 2003).

In this section we examine two alternative approaches to defining GHG trading ratios to reflect uncertainty in emissions inventories and maintain the environmental integrity of trades. As in Section 2.1, we start from the premise that any adjustments to the trading ratios should be designed so that allowance trading does not diminish environmental quality. There are at least two possible ways to interpret this:

- In the first situation, countries have emissions commitments, and a country is found to be in compliance with its commitment if its estimated emissions inventory is less than or equal to its commitment. In this case, the trading ratio is defined so that the upper bound of a confidence interval (say, 95%) around their estimated combined emissions is unchanged by trading, relative to a system of binding commitments (that are met). Thus, we can be confident that the upper bound of the uncertainty band around total combined emissions does not rise as a result of

trading. Note that we do not know whether estimated total combined emissions rise or fall.

- In the second situation, countries have emissions commitments, but these have been converted into what are referred to as *targets*; that is, country commitments are adjusted to reflect uncertainty in a manner similar to that in Section 2.1. A country is assumed to be in compliance if its emissions inventory is less than or equal to its *target*. In this case, the trading ratio is determined so that the probability that two countries exceed their aggregate (i.e., combined) emissions *commitment* is the same before and after trading. Thus, we want to be confident that actual combined emissions do not rise as a result of trading. Again, estimated total combined emissions may rise or fall.

Sections 3.1 and 3.2 below address each of these situations in turn. It turns out that, given reasonable assumptions about uncertainty and environmental goals, the intuition behind the nutrient trading program – that uncertain emissions should be less valuable than certain emissions – is not necessarily justifiable from an environmental and statistical perspective. Whether, and how, trading ratios should be adjusted to account for uncertainty depends, in fact, on the characteristics of the uncertainty estimate. In Section 3.3 we look at the characteristics that the uncertainty estimate would need to possess to be viable in these applications, building on the discussion in Section 2.2. We also discuss some additional issues in the practical application of trading ratios.

3.1 Trading Ratios: Upper Bound Emissions are Unchanged

The approach developed below ("upper bound") begins with the idea that we want to be confident that emissions do not rise as a result of trades. The starting point for this approach is the idea that, given an environmental goal, the purpose of both national commitments and the trading system is to ensure – with a reasonable amount of certainty – that this goal is not exceeded. In this case, the assumed "goal" is an upper bound of a probability distribution around mean estimated emissions. For example, the upper bound might be defined as the upper end of a 95% confidence interval around the mean; that is, the

---

[22] The impact of altered management practices at a farm, for example, will depend on the effectiveness of practices at the farm in reducing "edge of farm" nutrient discharges (which is highly site specific), on weather, on how spatially removed the farm is from an adjacent water body, and on conditions in adjacent receiving water (King & Kuch, 2003).

🖄 Springer

97.5 percentile value.[23] In this case, we can be very confident that actual emissions will not exceed the upper bound value, given the mean emissions estimate. Thus, one possible approach to designing a trading system is to define trading ratios such that trades *do not change the upper bound*. Thus, trading will not change the likelihood that we achieve our desired environmental goal (measured in terms of actual emissions), even if the mean emissions estimate changes.

Suppose there are two countries, A and B. Without loss of generality, choose emissions units so that Country B anticipates reducing emissions by one unit below its commitment. Country A has committed to reduce emissions to an amount A, and Country B has committed to reduce emissions to an amount B. Thus A and B are the emissions commitments of Countries A and B, respectively, divided by the anticipated additional emissions reduction by Country B. Suppose we have good information on the percentage (or fractional) uncertainty (denoted u) range associated with a 95% confidence interval for the emissions estimates for two countries. Thus, if B achieves its goal, its upper bound (97.5th percentile) emissions will equal $(1+u_B)B$. Similarly, if A achieves its goal, its upper bound emissions will equal $(1+u_A)A$.[24]

Suppose further that B anticipates reducing emissions below its commitment, and that A anticipates being unable to meet its commitment. The question then becomes, If B anticipates reducing emissions by one unit below its commitment, so that emissions in Country B equal $(B-1)$, by how much could Country A increase its emissions without violating the upper bound constraint? If the amount that Country A could

increase its emissions is called x, then x also gives the trading ratio between the two countries; one unit of emission reductions in Country B is worth x units of extra emissions in Country A. Thus, Country A will be willing to pay to B, for each unit B sells, an amount equal to the amount it would cost Country A to reduce emissions by x units.

Assuming approximate normality, the estimate of total emissions represented by the commitments has mean $A+B$ and adjusted standard deviation[25] given by

$$SD = \sqrt{u_A^2 A^2 + u_B^2 B^2},\qquad(1)$$

so that the upper bound for the total emissions represented by the commitments is given by

$$BOUND = A + B + SD.\qquad(2)$$

The post-trading total for the relevant sectors has mean $A + x + B - 1$ and upper bound

$$PBOUND = A + x + B - 1$$
$$+ \sqrt{u_A^2(A+x)^2 + u_B^2(B-1)^2}.\qquad(3)$$

A reasonable argument requires that trading should not change the upper bound (although the mean does change), so that we are just as confident as before of not exceeding the given upper bound. We therefore choose x to be the solution of BOUND = PBOUND. To solve this equation, first write it in the form

$$x - 1 = SD - \sqrt{u_A^2(A+x)^2 + u_B^2(B-1)^2}$$
$$= SD - SD2.\qquad(4)$$

Next, subtract SD from both sides and square the resulting equation to obtain

$$SD2^2 - SD^2 = (x-1)^2 - 2(SD)(x-1).\qquad(5)$$

---

[23] A "95% confidence interval" is an interval calculated from observational data such that the interval would be expected to include the unknown true value (e.g., total GHG emissions) for 95% of possible data sets, although we generally will not know whether or not this is true for a given data set. Since emissions inventory estimation often uses non-statistical methods (e.g., expert judgment) and methods not based on observational data, the term 95% confidence interval is here extended to mean any interval that in some sense is assumed to have a 0.95 probability of including the unknown true value. The upper bound is typically assumed to be the 97.5th percentile, and the lower bound the 2.5th percentile, so that the same 2.5% of the values lie above and below the confidence interval.

[24] For simplicity, we assume that the uncertainty (expressed as a percentage) is unchanged for the sector or country by activities that increase or decrease emissions.

[25] Strictly, this equation represents the standard deviation of the sum of emissions from A and B, multiplied by a scalar. The magnitude of the scalar (which may equal 1) depends on the width of the confidence interval for which the uncertainties are calculated and on the shape of the distribution of emissions. The scalar would equal 1.96 for a 95% confidence interval if emissions were normally distributed. It is assumed for this equation that the uncertainties represent the same level of confidence for both A and B.

This gives a quadratic equation for x

$$u_A^2(2Ax + x^2) + u_B^2(-2B + 1) = (x - 1)^2 - 2(SD)(x - 1), \text{ giving}$$

$$x = \frac{-\beta + \sqrt{\beta^2 - 4\alpha\chi}}{2\alpha}, \text{ where}$$
$$\alpha = u_A^2 - 1,$$
$$\beta = 2Au_A^2 + 2 + 2(SD),$$
$$\chi = u_B^2(1 - 2B) - 1 - 2(SD).$$

$$(6)$$

If A and B are large (relative to 1, the quantity of emissions to be sold), then the solution for x is approximately

$$x = \frac{SD + B \cdot u_B^2}{SD + A \cdot u_A^2}. \tag{7}$$

(This can be shown using Taylor series expansions. Note that x is a dimensionless ratio, as are the values of SD, A, and B, since everything is relative to the additional one unit emissions reduction by Country B.) Thus, unless a country is selling or buying a large portion of its emissions, the simpler equation is a reasonable approximation of the trading ratio. In this equation, x could be bigger or smaller than 1 depending on the relative sizes of the means (A and B) and of the uncertainties (u_A and u_B).

The equation for x above illustrates that the trading ratio that satisfies this approach is not simply a function of the uncertainty of each country's inventory. Rather, the trading ratio depends on (1) the magnitude of estimated emissions in Countries A and B; (2) the absolute uncertainty (i.e., the standard deviation) of total emissions from the two countries; and (3) the relative uncertainties surrounding emissions in Countries A and B.

In particular, consider the numerator of the equation defining x in Eq. 7. All else being equal, x will be higher if the second term, $B u_B^2$, is greater. In other words, if any three out of A, B, u_A, u_B, and SD are held fixed, then x increases with the second term, $B u_B^2$, because of the bound condition BOUND = PBOUND.

This second term combines the uncertainty of Country B's emissions with the magnitude of its emissions. (Formally, it is proportional to the variance in B's emissions divided by B's mean estimated emissions.) Thus, as this term rises, Country A should (from a global perspective) pay *more* to reduce emissions from B, or, equivalently, emission reductions purchased from B should translate into fewer emissions by Country A. The rationale is that emission reductions in Country B contribute more to reducing

the upper bound of the combined emissions from A and B than would the same quantity of emission reductions by Country A. Similarly, as the analogous term for Country A (the second term in the denominator) rises, Country A should pay *less* (from a global perspective) for emission reductions from Country B, because the emission reductions from Country A would do more to reduce the upper bound than would emission reductions from Country B.[26]

A simple example may help clarify how the equation for x of Eq. 7. would work. Suppose that Countries A and B have both committed to emissions of 100 t. The uncertainty in the emissions estimate is 40% for Country A and 50% for Country B.[27] Country B finds that it is cheaper to reduce its emissions than it anticipated, and Country A finds that it is more difficult to meet its commitment than anticipated. Thus, Country A finds that its estimated emissions inventory equals 110 t, and it needs to purchase 10 t of emission reductions (emission allowances) from another country. Country B has estimated emissions of 90 t, and so it has 10 t to sell. Using the above equation, x equals 1.11. Country A then purchases about 9 t of Country B's excess reductions to offset A's excess of 10 t of emissions, and so meet its own commitments. Note that, whenever x is greater than 1, estimated total emissions between the countries will *rise* as a result of the trade. This and other examples are illustrated in Table 3.

The trading ratio formula may seem counter to expectations, because it implies that the emissions with the greatest uncertainty are the *most* valuable to buyers. The intuitive explanation is that if Country A has relatively certain emissions and Country B has relatively uncertain emissions, then A's contribution to the overall upper bound (to $A+B$) is small compared with the reduction in the upper bound caused by a one-unit reduction in B's emissions. Effectively, Country A is given a bonus because each reduction in B's uncertain emissions is being swapped for more certain emissions from A. We value reductions in uncertain sources more highly because such reductions essentially begin to remove the emissions from uncertain source categories

---

[26] The impact of the SD term depends on the ratio of the uncertainty products.

[27] While this large uncertainty between countries is unlikely for developed countries, it is certainly possible for trades between source categories. Moreover, the large uncertainty serves to illustrate the workings of the trading ratio.

Table 3 Illustrative trading ratios

Country A (buyer)		Country B (seller)		X (trading ratio)
Emissions commitment (t)	Uncertainty (%)	Emissions commitment (t)	Uncertainty (%)	
100	40	50	5	0.72
100	40	50	20	0.76
100	5	50	40	1.37
100	20	50	40	1.12
100	30	50	40	0.98
50	40	100	5	0.73
50	40	100	20	0.89
50	5	100	40	1.39
50	20	100	40	1.32
50	30	100	40	1.24
100	40	100	10	0.74
100	40	100	30	0.89
100	40	100	50	1.11

from the inventory and from the environmental system; for a given emissions estimate, the environment would be better off if those emissions came only from the most certain source categories, because then we would have the best idea of what emissions really look like. This does not (as some suppose) argue for removing uncertain emissions from the *inventory*, but rather places a higher premium on removing more uncertain emissions from the *environment*.

In Section 3.2, we explore a variant of this approach, using a slightly different environmental goal. In Section 3.3 we return to the question of whether this approach makes sense from the perspective of the uncertainty characteristics of the GHG inventory, and discuss some possible implications for nutrient trading as well.

3.2 Trading Ratios: Probabilities of Exceeding Emissions Commitments are Unchanged

An alternative trading ratio can be developed based on limiting the probability of exceeding the emissions commitment (i.e., combining some of the ideas in Sections 2.1 and 3.1). Suppose Countries A and B have emissions commitments under the Kyoto Protocol of E_A and E_B, respectively. Instead of adjusting inventory estimates to reflect uncertainty (as in Section 3.1), each country is assumed to have an emissions target A or B, where the target is derived by

adjusting the emissions commitment level to reflect uncertainty. Specifically, the target is determined so that if a country has an estimated emissions inventory that equals the emission target, then the probability is 95% that actual emissions do not exceed the emissions commitment by more than 10% (for that country). As in Section 3.1, we choose the emissions units such that Country A wants to buy one unit of emissions from Country B. Instead of defining trading ratios to preserve the upper bound (as in Section 3.1), this here we define trading ratios to preserve the probability that total estimated emissions from the two countries sum to less than their combined emissions commitments.

Let the two countries have fractional uncertainties u_A and u_B. Assume that the emissions targets are defined so that, at the targets, the probability of not exceeding the emissions commitment by more than 10% is 95%. Assume that emissions are (approximately) normally distributed. For Country A, with estimated emissions meeting the adjusted target (A), the 95% confidence interval for actual emissions is

$$\text{Emissions} = A \pm Au_A = A \pm 1.96 \text{ Std. Dev. (emissions).} \tag{8}$$

Thus the probability of not exceeding the emissions commitment by more than 10% equals

$$\Phi\left(\frac{E_A(1.1) - A}{Au_A/1.96}\right) = 0.95, \tag{9}$$

where Φ denotes the cumulative distribution function of a standard normal random variable. This equation has the solution

$$1.1\,E_A = A\left(1 + \frac{1.64}{1.96}u_A\right). \tag{10}$$

A similar equation applies for Country B.

Before trading, the probability that the estimated combined emissions for the two countries will not exceed the combined emissions commitment by more than 10% equals

$$\Phi\left(\frac{(E_A + E_B)(1.1) - (A + B)}{\sqrt{A^2 u_A^2 + B^2 u_B^2}\big/1.96}\right). \tag{11}$$

If B sells one unit and A is allowed x units for that trade, then the mean combined emissions after trading will be $A + x + B - 1$, and the standard deviation will

houses. However, such a system could be controversial if it is perceived as inequitable. In particular, the trading ratio at which a country sells emissions to the clearinghouse will be different for each country (as well as for each source and gas, if that is the level at which trading ratios are calculated), so that a unit of one country's emissions may be more valuable than a unit sold by another country. Similar issues arise for the different buying ratios.

3.3.3 Implications for Trading Ratios in Practice

The discussion in Section 3 has followed a systematic approach of defining an environmental goal and then identifying the statistical implications of that goal. The discussion suggests that, as in the discussion of the adjustment factor in Section 2, there is no unique method for calculating trading ratios, but rather the appropriate ratio depends on the environmental goal. Moreover, for the weak environmental goal examined here, the conventional wisdom (that uncertain emissions should be valued less in a trade than certain emissions) is not borne out. Rather, the trading ratio depends on both the uncertainty and the magnitude of emissions. Further, because uncertain emissions contribute more to increasing the upper bound on the emissions estimate than do certain emissions, reductions in uncertain emissions tend to be valued more highly than reductions in certain emissions (given the definition examined here). Consequently, we should not *assume* that a trading ratio less than 1 (i.e., that a one-unit reduction in uncertain emissions offsets less than one unit of increased certain emissions) is necessitated by the uncertainty of emission reductions.

The trading ratio developed here assumes that we have valid estimates of statistical uncertainty and that we believe that our measures of statistical uncertainty adequately capture and represent all significant sources of uncertainty. Thus, the requisite characteristics of an uncertainty estimate described in Section 2 – for example, that it be objective and verifiable – are even more crucial to a trading ratio. In addition, because a trading ratio that includes an adjustment for country- or source-specific uncertainty involves calculating the ratio each time a trade is made, the system could be administratively intractable or at least very costly to participate in and administer. One solution might be to develop a clearinghouse so

that trades occur only through the central authority, and so bilateral trades are not examined on a case-by-case, individual basis.

4 Uncertainty Analysis as a Tool for Inventory Improvement

In the context of national GHG inventories, the process of producing an uncertainty analysis can be divided into four parts: (1) the rigorous investigation of the likely causes of data uncertainty and quality; (2) the creation of quantitative uncertainty estimates and parameter correlations; (3) the mathematical combination of those estimates when used as inputs to a statistical model (e.g., first-order error propagation or Monte Carlo method); and (4) the selection of inventory improvement actions to take in response to the results of the uncertainty analysis. There has been a tendency in much uncertainty work associated with national GHG inventories to focus on the second and third parts, with less effort expended on the first and fourth.

Although the process of modeling the interactions between the uncertainties in parameter values can be instructive, in isolation it does not provide the type of specific information needed to isolate the causes of data quality problems so that they can be corrected or lessened. We refer to any approach to uncertainty analysis that puts an intense focus on the first and fourth parts of this process as *investigation focused*. An investigation-focused approach to uncertainty analysis can both provide the kind of rigorous information needed to more credibly quantify the uncertainties in parameters for use in modeling and simultaneously lead to a system focused on achieving real data and inventory quality improvements.

An investigation-focused approach to uncertainty analysis requires that inventory developers work closely with data suppliers and researchers to (1) exchange information on the inventory's data quality requirements and actual data collection practices; (2) identify activity data reporting or collection problems; (3) identify situations where there is a lack of empirical data for emission factors or other parameters; (4) identify situations where the variability in an inventory parameter is high; (5) identify situations where there is a lack of scientific consensus of the appropriate estimation method for an inventory parameter or category; and (6) identify specific actions that can be taken to correct or mitigate each problem.

The process of analyzing uncertainties can provide a systematic approach for the thorough investigation of the data underlying an inventory and a basis for a more formal understanding of data quality. By jointly identifying specific causes of uncertainty and approximating the magnitude of their effect on data quality, inventory practitioners and data collection agencies can generate better quantitative uncertainty estimates and hopefully also produce better arguments for investments in data quality improvements (e.g., expanded data collection or more research).

This process of implementing an uncertainty analysis effort that is *investigation focused* has been found to be helpful to the authors in the process of preparing inventories at an individual facility (i.e., project), for a corporation, and at the national level. These benefits of this type of approach can be summarized as follows:

- Promoting a broader learning and quality feedback process within the national inventory process.
- Supporting efforts to qualitatively understand and document the causes of uncertainty and help identify ways of improving inventory quality. For example, collecting the information needed to determine the statistical properties of activity data and emission factors forces researchers to ask hard questions and to carefully and systematically investigate data quality.
- Establishing lines of communication and feedback with national statistical agencies, researchers, and other data suppliers, in order to identify specific opportunities to improve the quality of the data and methods used.
- Providing valuable information to reviewers, stakeholders, and policy makers for setting priorities for investments aimed at improving data sources and methodologies.
- Informing policy makers engaged in negotiating future climate change treaties regarding the possible range of confidence they can have in the monitoring of future targets.

It should be obvious that an investigation-focused approach to uncertainty is one that should be tightly integrated with an inventory agency's quality control and quality assurance (QA/QC) processes. In many ways, an investigation-focused approach to uncertainty is simply a more in-depth approach to quality management in that it is a process to rigorously identify the causes of data quality problems, especially ones that the general quality control processes already in place in a country are unlikely to catch. These problems will often involve issues of incomplete data or other systematic biases in the data, which also happen to be key issues for developing a quantitative uncertainty analysis.

An investigation-focused uncertainty analysis can be performed solely on a qualitative basis and still provide useful information for inventory improvements. However, it can provide more useful information for prioritizing the allocation of scarce resources to inventory improvements if it also produces rough quantitative uncertainty estimates. These rough quantitative uncertainty estimates can then be combined with estimates of how much each data quality improvement investment is expected to lower the uncertainty in a particular parameter.

The required characteristics of quantitative uncertainty estimates are obviously less strict if they are only to be used as input for deciding how to prioritize inventory improvements than if they are to be used for a particular policy purpose. For example, it is less critical that rigorous expert elicitation protocols be utilized to increase the comparability of uncertainty estimates across parameters, source categories, and countries. Moreover, because with an investigation-focused approach quantitative uncertainty estimates are only used internally by an inventory agency for allocating resources, the manipulation (i.e., gaming) of uncertainty estimates for the benefit of a particular party is less of a concern. However, particular experts engaged in inventory work within a country may still have an incentive to exaggerate the magnitude of particular uncertainties or the benefits of particular actions in terms of lowering that uncertainty in order to obtain greater budget allocations.

In summary, the purpose of an investigation-focused approach to uncertainty analysis is to improve inventory quality, not just to assess inventory uncertainty. Inventory agencies do not have to choose between an investigation-focused and Monte Carlo-type uncertainty analysis. The former should be seen as a way of obtaining better results than can be obtained from the latter. However, for an inventory agency with limited resources for uncertainty analysis, the quality of its inventory will likely benefit the most if those resources are shifted to the first and fourth parts of the process. Instead of expending resources on quantification and developing models to combine subjective (i.e., expert-judgment-based) estimates, limited resources can be expended on identifying and correcting real data quality problems.

5 Conclusions

Information on the uncertainties in a national GHG inventory – including quantitative estimates of uncertainty – can have a variety of different applications that in turn can satisfy a variety of different goals. For uncertainty information to have practical applications, however, it needs to have characteristics that match the application. These characteristics are particularly restrictive for applications of quantitative inventory uncertainty estimates for policy purposes, such as adjusting emissions for determining compliance.

Consider, for example, a policy that involves an adjustment to an inventory or an emissions trading ratio that is designed to capture uncertainty. Such an adjustment mechanism can, at a minimum, be evaluated against the same types of criteria that we would require of other environmental policies, such as cost-effectiveness, fairness, and administrative feasibility, among others. In turn, these criteria suggest key characteristics that an uncertainty estimate should have if it is to be the basis for an adjustment mechanism, namely, (1) it should be comparable across countries; (2) it should be relatively objective, or at least subject to review and verification; (3) it should not be subject to gaming by countries acting in their own self-interest; (4) it should be administratively feasible to estimate and use; (5) the quality of the inventory uncertainty estimate should be high enough to warrant the additional compliance costs its use in an adjustment factor may impose on countries; and (6) in order to fully secure environmental benefits, it should attempt to address all types of inventory uncertainty, particularly in the case of trading ratios.

In the context of the current state of national GHG inventories, uncertainty estimates do not have the characteristics outlined above. For example, the information used to develop quantitative uncertainty estimates for national inventories is quite often based on expert judgments, which are, by definition, subjective rather than objective. These expert judgments do not undergo any rigorous type of review or verification and are unlikely to be comparable across countries, source and sink categories, parameters, and time, because of differences across the experts producing the judgments.

Over time, however, the authors hope that uncertainty estimates will come closer to possessing these characteristics. As national inventories improve, so should our ability to (1) objectively estimate uncertainties (i.e., by linking uncertainty estimates to specific measurement techniques); (2) review country-specific uncertainty estimates; and (3) elaborate detailed guidance for conducting uncertainty analyses. Whether country-specific quantitative uncertainty estimates of national GHG inventories will ever be "good enough" to base adjustment policies on is highly debatable and depends not only on having the political will to accomplish these changes, but also on the potential technical limits in uncertainty analysis.

Assuming that we can develop quantitative uncertainty estimates for GHG inventories with the characteristics necessary to apply them to policy applications, policy makers still must design an appropriate adjustment mechanism. In turn, the appropriate design of inventory adjustments or trading ratios depends, at least in part, on what type of adjustment is statistically valid; this in turn depends on how the policy goal is defined. Consequently, the design of adjustment mechanisms can benefit from a systematic approach in which policy makers (1) identify clear environmental goals; (2) define these goals precisely in terms of relationships among important variables (such as emissions estimate, commitment level, or statistical confidence); and (3) develop quantifiable adjustment mechanisms that reflect these environmental goals as they are defined. In some cases, a systematic approach may suggest that the statistically valid approach is not the one that is commonly accepted by the conventional wisdom.

An investigation-focused (i.e., qualitative) uncertainty analysis can (1) provide the type of information necessary to develop more substantive, and potentially useful, quantitative uncertainty estimates-regardless of whether those quantitative estimates are used for policy purposes-and (2) provide information needed to understand the likely causes of uncertainty in inventory data and thereby point to ways to improve inventory quality (i.e., accuracy, transparency, completeness, and consistency). Too often, analysts simply assume that uncertainty estimation will provide quality improvements, rather than structuring a process of investigation, analysis, and feedback that is designed to obtain real quality benefits.

Implementing a process of investigating the uncertainty of the emissions inventory may require resolving potentially competing priorities. A process that is intended to derive quantitative uncertainty estimates should involve a different emphasis than a process that is focused on producing inventory improvements. Similarly, deriving uncertainty estimates for use in a policy context may require a very different emphasis than if the estimates are for use in scientific or modeling applications. This paper has begun to explore these issues by identifying how the expected use or application of the uncertainty estimates influences the characteristics that the uncertainty estimate should have. The focus in the paper is on two particular uses: policy (adjustment schemes for emissions inventories and for trading ratios) and inventory improvement. We find that, indeed, identifying the application or applications of the results of an uncertainty analysis is critical to how it should be designed and implemented.

Acknowledgement Portions of this work were supported by Environment Canada and the US Environmental Protection Agency. The views expressed herein are entirely those of the authors.

References

Bartoszczuk, P., & Horabik, J. (2007). Tradable permit system: Considering uncertainty in emission estimates. *Water, Air, & Soil Pollution: Focus* (in press) doi:10.1007/s11267-006-9110-x

Cohen, J., Sussman, F., & Jayaraman, K. (1998). *Improving Greenhouse Gas Emission Verification*, Final report prepared by ICF Consulting and submitted to Environment Canada, 2 January. Subsequently released as an Environment Canada Report, February.

ISO (1993). *International vocabulary of basic and general terms in metrology*, Second edition. Geneva, Switzerland: ISO (International Organization for Standardization).

Jonas, M., & Nilsson, S. (2007). Prior to economic treatment of emissions and their uncertainties under the Kyoto Protocol: Scientific uncertainties that must be kept in mind. *Water, Air, & Soil Pollution: Focus* (in press) doi:10.1007/s11267-006-9113-7

Kasa, K. (2000). Knightian uncertainty and home bias. *Federal Reserve Bank of San Francisco Economic Letter.* (October 6)

King, D. M., & Kuch, P. J. (2003). Will nutrient trading ever work? An assessment of supply and demand problems and institutional obstacles. 33 *ELR* 10352, Environmental Law Institute, Washington, DC, USA.

Knight, F. H. (1921). *A treatise on probability.* London, UK: Macmillan.

Monni, S., Syri, S., Pipatti, R., & Savolainen, I. (2007). Extension of EU emissions trading scheme to other sectors and gases: Consequences for uncertainty of total tradable amount. *Water, Air, & Soil Pollution: Focus* (in press) doi:10.1007/s11267-006-9111-9

Morgan, M. G., & Henrion, M. (1990). *Uncertainty: A guide to dealing with uncertainty in quantitative risk and policy analysis.* Cambridge, UK: Cambridge University Press.

Nahorski, Z., Horabik, J., & Jonas, M. (2007). Compliance and emissions trading under the Kyoto Protocol: Rules for uncertain inventories. *Water, Air, & Soil Pollution: Focus* (in press) doi:10.1007/s11267-006-9112-8

Nilsson, S., Shvidenko, A., Jonas, M., & McCallum, I. (2007). Uncertainties of a regional terrestrial biota full carbon account: A systems analysis. *Water, Air, & Soil Pollution: Focus* (in press) doi:10.1007/s11267-006-9119-1

Nishimura, K. G., & Ozaki, H. (2001). Search and Knightian uncertainty. *Working paper*, University of Tokyo, Japan.

Rousse, O., & Sévi, B. (2007). The impact of uncertainty on banking behavior: Evidence from the US sulfur dioxide emissions allowance trading program. *Water, Air, & Soil Pollution: Focus* (in press) doi:10.1007/s11267-006-9109-3

Rypdal, K., & Winiwarter, W. (2000). Uncertainties in greenhouse gas emission inventories – evaluation, comparability and implications. *Environmental Science and Policy*, 107–116.

Sussman, F. (1998). Compliance and uncertainty in emissions inventories. Presentation to UNCTAD meeting on verification and accountability, London, UK, 6 April.

Sussman, F., Cohen, J., & Jayaraman, K. R. (1998). Uncertain emissions inventories, compliance, and trading. Presentation at Global Climate Change: Science, Policy, and Mitigation/Adaptation Strategies, Annual meeting of the Air & Waste Management Association, Washington, D.C., USA 13–15 October.

Taylor, J. R. (1997). *An introduction to error analysis: The study of uncertainties in physical measurement*, Second edition. Sausalito, CA, USA: University Science Books.

UNFCCC (2000). Views from parties on national systems, adjustments and guidelines under Articles 5, 7 and 8 of the Kyoto Protocol. *FCCC/SBSTA/2000/MISC.1,* 24 February.

Webster, M., Forest, C., Reilly, J., Babiker, M., Kicklighter, D., Mayer, M., et al. (2003). Uncertainty analysis of climate change and policy response. *Climate Change, 61,* 295–320.

Winiwarter, W. (2007). National greenhouse gas inventories: Understanding uncertainties versus potential for improving reliability. *Water, Air, & Soil Pollution: Focus* (in press) doi:10.1007/s11267-006-9117-3

Winiwarter, W., & Rypdal, K. (2001). Assessing the uncertainty associated with national greenhouse gas emission inventories: A case study for Austria. *Atmospheric Environment,* 5425–5440.

Water Air Soil Pollut: Focus (2007) 7:475–482
DOI 10.1007/s11267-006-9115-5

Modeling Afforestation and the Underlying Uncertainties

M. Gusti

Received: 27 April 2005 / Accepted: 12 March 2006 / Published online: 19 January 2007
© Springer Science + Business Media B.V. 2007

Abstract A dynamic model of the carbon budget of an oak forest ecosystem that takes into account forest stand age was developed. A numerical experiment was designed to simulate the afforestation process, and a Monte Carlo simulation was performed to determine how parameter uncertainties and environmental variability influence the result. It was found that while the total amount of carbon stored in the ecosystem increases from 1.9 kg C/m^2 to 4.4 kg C/m^2 over the following 20 years, the relative standard deviation increases from 9 to 21%. The contribution of varying climate and carbon dioxide parameters to total uncertainty is substantial; for example, the standard deviation at the 10th modeling year for phytomass doubles and the uncertainties of the soil pool and total accumulated carbon increase by a factor of nearly 1.4, while the uncertainty of the litter pool stays almost at the same level.

Keywords afforestation · mathematical model · Monte Carlo simulation · uncertainty estimation

M. Gusti (✉)
International Institute for Applied Systems Analysis,
A-2361 Laxenburg, Austria
e-mail: gusti@iiasa.ac.at

1 Introduction

The Kyoto Protocol to the United Nations Framework Convention on Climate Change commits most developed countries to reducing their greenhouse gas emissions. The Protocol also defines some "legal" means that can be used by countries to reach the required emission levels, one of these being afforestation (the planting of forests on land where forests have not grown for the last 50 years); afforestation is a "natural" way of trapping the atmospheric carbon dioxide (CO_2) in long-living phytomass, detritus, and humus. Countries can use the carbon credits accumulated through afforestation activities (1) to fulfill their obligations under the Kyoto Protocol; and (2) to participate in the market-based mechanisms created under the Protocol (i.e., international emissions trading, joint implementation, and the clean development mechanism).

The uncertainty regarding emission estimates plays an important role in emissions trading (see, in particular, Bartoszczuk & Horabik, 2007; Monni, Syri, Pipatti, & Savolainen, 2007; Nahorski, Horabik, & Jonas, 2007). While, in emissions trading between EU15 members, land use change and forestry do not influence total uncertainty (Monni et al., 2007), for some countries the effects of afforestation, reforestation, and deforestation can be considerable.

When developing an afforestation project, it is important to estimate the amount of carbon that can be accumulated in the ecosystem during a given period

 Springer

of time, bearing in mind the uncertainty of the estimates and the risk of not achieving the desired result. The point is that the forest will grow in a changing environment. Prognostic modeling can help in obtaining a first guess as to the values of accumulated carbon and will demonstrate not only the uncertainties and risks but also the influence of environmental variability.

To account for forest stand age, we propose the use of a dynamic mathematical model of the carbon budget of an oak forest ecosystem (discussed in Gusti, Bun, Dachuk, & Shpakivska, 2004), which incorporates growth functions (Shvidenko, Venevsky, Raile, & Nilsson, 1996) and regression expressions (Lakida, Nilsson, & Shvidenko, 1996). To describe phenology, a function of the monthly mean temperature is developed. To estimate the available water in the ecosystem, a simple mathematical model, which accounts for the effects of frozen water accumulation in winter and the thawing of ice in spring, is elaborated. The uncertainties of the model parameters (including temperature, precipitation, and atmospheric carbon dioxide) are modeled with random generators.

The study is illustrative, as information about the model parameters is incomplete; thus, uncertainty classes of 10% and 20% of the relative standard deviation, as well as assumptions on probability distribution types (normal or uniform), were introduced. However, many other factors influencing the forest carbon budget (e.g., insects and fires) were not taken into account.

2 Description of the Model and Experiment

2.1 Description of the Model

In the mathematical model of the carbon budget of an oak forest the following carbon pools are considered: phytomass (leaves, distinguished using a regression expression), litter (five reservoirs: foliage, stems, branches, coarse roots, and fine roots), and soil organic matter. The following carbon flows are also considered: atmosphere–phytomass, phytomass–litter (litter sorted into five types using regression expressions), litter–atmosphere, litter–soil, soil–atmosphere, and phytomass–boundary of the ecosystem (harvested phytomass).

The mathematical model of the carbon budget is presented in the form of a system of ordinary differential equations of the first order:

$$\frac{dX_{ph}}{dt} = v_{ap} - \left(v_{plf} + v_{pls} + v_{plb} + v_{plcr} + v_{plfr} + v_{ph}\right),$$

$$\frac{dX_{lf}}{dt} = v_{plf} + v_{hlf} - \left(v_{lfa} + v_{lfs}\right),$$

$$\frac{dX_{ls}}{dt} = v_{pls} + v_{hlb} - \left(v_{lsa} + v_{lss}\right),$$

$$\frac{dX_{lb}}{dt} = v_{pbl} + v_{hlb} - \left(v_{lba} + v_{lbs}\right),$$

$$\frac{dX_{lcr}}{dt} = v_{plcr} + v_{hlcr} - \left(v_{lcra} + v_{lcrs}\right),$$

$$\frac{dX_{lfr}}{dt} = v_{plfr} + v_{hlfr} - \left(v_{lfra} + v_{lfrs}\right),$$

$$\frac{dX_s}{dt} = v_{lfs} + v_{lss} + v_{lbs} + v_{lcrs} + v_{lfrs} - \left(v_{sa} + v_{saq}\right),$$

where X with subscripts denotes carbon pools (kg C/m^2; ph is phytomass, lf is foliage litter, ls is stem and branch litter (diameter >10 cm), lb is branch litter (diameter <10 cm), lcr is coarse root litter, lfr is fine root litter), and v with subscripts denotes carbon flows between corresponding reservoirs (kg C/(m^{2}year)), for example, ap is atmosphere–phytomass, plf is phytomass–foliage litter, ph is phytomass–harvested phytomass, and saq is soil–aquatic system.

Intensity of net photosynthesis (the v_{ap} flow) is presented with a complex function:

$$v_{ap} = \alpha_{ap} * F_l * \min\left\{F_f, F_c, F_w\right\},$$

where α_{ap} is the calibration coefficient, F_l is the function of the mass of leaves, which, in turn, is a function of forest stand age (A, years), F_T is dependence on the monthly air temperature (T, °C), F_c is dependence on the monthly concentration of atmospheric CO_2 (C, ppmv), and F_w is dependence on the monthly amount of available water (W_a, kg/m^2).

Let us now consider the main functions. The mass of leaves (denoted as f) is defined with a regression equation (see Equation for R below), but the time at which leaves appear in oak forests is controlled by the air temperature (T_{lg}, °C). The process is described with the expression:

$$F_l = \frac{1}{1 + \exp\left(0.9 * \left(-T + T_{lg}\right)\right)} * \frac{R_f * X_{ph}}{R_{tot}}.$$

The functions F_T, F_c, and F_w are defined in Gusti (2002). The optimal temperature for photosynthesis is chosen to be equal to the normal temperature in July in a corresponding vegetation belt, which is 18.4°C for oak forests.

The flow of phytomass–foliage litter is denoted by v_{plf}. The time of leaf fall is controlled by the air temperature, which is decreasing (T_{lfb} is the temperature of the mass fall; T_{lfe} is the temperature when the leaves stop falling, measured in °C), but the intensity is controlled by the mass of leaves:

$$v_{plf} = \begin{cases} 15 * \left(\dfrac{1}{1 + \exp\left(1.2*(T - T_{lfe})\right)} \right. \\ \qquad \left. - \dfrac{1}{1 + \exp\left(1.2*(T - T_{lfb})\right)} \right) \\ \qquad * \dfrac{R_f * X_{ph}}{R_{tot}}, \quad \text{if } \dfrac{dT}{dt} < 0. \\ \\ 0, \qquad \text{otherwise} \end{cases}$$

The monthly mean air temperatures of major phenology changes in the ecosystems of oak forests are compiled using observations made in the Roztochchya nature reserve in the Lviv region of Ukraine. The temperature when leaves appear (T_{lg}) is 12.5°C; the temperature at mass leaf fall (T_{lfb}) is 9.0°C; and the temperature at which leaves falling stop (T_{lfe}) is 5.7°C.

Phytomass is divided into fractions (stems and branches with a diameter >10 cm are denoted as s; branches with a diameter <10 cm are denoted as b; coarse roots are denoted as cr, and fine roots are denoted as fr) using a regression expression (Karjalainen & Liski, 1997):

$$R_i = a_0^i * A^{a_1^i},$$

where a_0 and a_1 are regression coefficient values as listed in Lakida et al. (1996); and units of R_i are t/m^3). The following relations between the regression expressions are also used: $R_{br} = R_{kr} - R_f, R_{fr} = R_f, R_{cr} = R_{bl} - R_{fr}, R_s = R_{ab} - R_{kr}$.

Carbon flows from phytomass to the corresponding litter reservoirs are defined with the following expression:

$$v_{pli} = \alpha_{pli} * \left(dM * R_i * \frac{X_{ph}}{GS * R_{tot}} + \frac{X_{ph} * R_i}{Turn_i * R_{tot}} \right), \ i$$

$$= \{s, b, cr, fr\},$$

where α_i is the calibration coefficient, dM is natural mortality (m^3/[ha year]), and GS is the growing stock

(m^3/ha) of the forest stand defined in Shvidenko et al. (1996) (we consider a forest of III-rd site index using the Orlov scale), and $Turn$ is the turnover time (years) for tree parts listed in Karjalainen and Liski (1997). For convenience, we assume $Turn = \infty$ for stem.

Litter mineralization is described with the following expression:

$$v_{lia} = kle * F_{phi} * F_{Tl} * F_{Pl} * X_i, \quad i = \{f, s, b, cr, fr\},$$

where kle is a calibration coefficient (year^{-1}) and F_{phi} is a function of phytomass amount (Kurtz, Apps, Webb, & McNamee, 1992):

$$F_{phi} = k_i + 0.5 * k_i \exp\left(-\frac{9.21 * X_{ph}}{0.065 * GS} \right),$$

$k = 0.045 \div 0.42$ (Karjalainen & Liski, 1997). F_{Tl} is a function of temperature (defined in Krapivin, Svirezhev, Yu, & Tarko, 1982), the parameter Q_{10} equals 2.25 (Shpakivska & Maryskevych, 2003). F_{Pl} is a function of the amount of available water:

$$F_{Pl} = 1 - \exp\left(-0.017 * W_a \right).$$

Mineralization of soil organic matter (V_{sa}) is determined with a similar expression, but different parameters are used, in particular, $Q_{10} = 1.84$ (Shpakivska & Maryskevych, 2003), and $F_{ph} = 1$.

Litter humification is: $v_{lis} = p * V_{lia}$, $i = \{f, s, b, cr, fr\}$ ($p = 0.19$). One assumes that the branches, roots, and stumps of harvested trees are left in the forest: $v_{hli} = R_i * Harv$, $i = \{f, s, b, cr, fr\}$; $V_{hls} = 0.3 * Harv$, where $Harv$ is the amount of harvested stem wood (kg C/(m^{2}year)). The flow $vsqa$ is introduced to take into account the increasing soil runoff when forests are harvested. In the current version of the model, the flow is set constant at 0.0004 kgC/(m^{2}year).

2.2 Submodel of Available Water in an Ecosystem

Let us now describe the dynamics of snow with the following equation:

$$\frac{dm_s}{dt} = -\left(v_a + v_p + v_w \right) * \left(1 - \exp\left(-2 * m_s \right) \right) * \gamma,$$

where v_a is melted snow (kg/(m^{2}year)) caused by the exchange of heat between snow and air (the equation

Table 1 Comparison of measured and modeled phytomass and net increment of an oak forest

		33	54	75	106	Root-mean-square error, %
	Age, years					
Phytomass, kgC/m^2	Measured	5.40	8.70*	11.56*	13.67*	16
	Modeled	5.40	10.05	12.99	15.58	
	Difference	0.00	1.35	1.43	1.91	
Net increment, kgC/(m^{2}year)	Measured	0.36	0.17	0.16	0.11	24
	Modeled	0.34	0.16	0.12	0.05	
	Difference	0.02	0.01	0.04	0.06	

*Values reduced to forest stand stocking 0.79

of ideal gas energy is used):

$$\nu_a = \begin{cases} 3/2 * (m_{air}/\mu) * R * (T + 273)/\lambda, \\ \quad if\ T > 0\ and\ m_s > 0, \\ 0, \quad if\ T < 0\ or\ m_s = 0 \end{cases}$$

where m_{air} denotes the mass of air near the ground (kg/m^2); μ is the molar mass of air (0.029 kg/mol); T is air temperature (°C); and λ is the specific heat of melting ice (3.34 10^5 J/kg).

ν_p denotes melted snow (kg/(m^{2}year)) caused by the exchange of heat between snow and rain water as well as the kinetic energy of rain:

$$\nu_p = \begin{cases} \{C_w * W_i * (T + 278) + 0.5 * W_i * v^2\}/\lambda, \\ \quad if\ T > 0\ and\ m_s > 0 \\ 0, \quad if\ T < 0\ or\ m_s = 0 \end{cases}$$

where C_w is the specific heat of water (4.3 10^3 J/(kg K)); W_i is precipitation (kg/m^2); and v is the mean vertical speed of rain droplets near the earth's surface (6.5 m/s) (Helming, 2001).

ν_{SR} is melted snow (kg/(m^{2}year)) caused by heat from solar radiation:

$$\nu_{SR} = \begin{cases} SR * (1 - \alpha)/\lambda, \ if\ T > 0\ and\ m_s > 0 \\ 0, \quad if\ T < 0\ or\ m_s = 0 \end{cases}$$

where SR is solar radiation W/(m^{2}year); and α is forest albedo (0.15). ν_{wt} is snow that has been weathered and blown by the wind: $\nu_{wt} = 0, 1 * m_s$. There is also a dimension factor, namely, $\gamma = 1\ s^{-1}$.

The amount of water (kg/m^2) available in the ecosystem and influencing the intensity of the carbon

cycle processes W_a is defined with the following expression:

$$W_a = \begin{cases} W_i + v_a + v_p + v_{SR} - v_{wt}, & if\ T > 0\ and\ m_s > 0 \\ 0, & if\ T < 0 \\ W_i, & if\ T > 0\ and\ m_s = 0 \end{cases}.$$

2.3 Calibration and Testing of the Model

For the calibration and testing of the oak forest model, the measurement data from four test plots of oak forests of different ages (33, 54, 75, and 106 years) are used (Borsuk et al., 1982). The root-mean-square error (RMSE) of phytomass modeling is 16% and the phytomass net increment is 24% (Table 1). During calibration only, the calibration coefficients (α_{ap}, α_{pli}, and kle) were tuned to minimize RMSE in all measured points.

The dynamics of the model after calibration was tested visually for plausibility in all trajectory points.

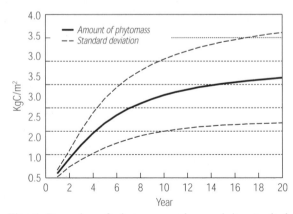

Fig. 1 Dynamics of phytomass carbon and its standard deviation

Water Air Soil Pollut: Focus (2007) 7:483–494
DOI 10.1007/s11267-006-9116-4

Spatial GHG Inventory: Analysis of Uncertainty Sources. A Case Study for Ukraine

R. Bun · M. Gusti · L. Kujii · O. Tokar ·
Y. Tsybrivskyy · A. Bun

Received: 27 April 2005 / Accepted: 12 March 2006 / Published online: 24 January 2007
© Springer Science + Business Media B.V. 2007

Abstract A geoinformation technology for creating spatially distributed greenhouse gas inventories based on a methodology provided by the Intergovernmental Panel on Climate Change and special software linking input data, inventory models, and a means for visualization are proposed. This technology opens up new possibilities for qualitative and quantitative spatially distributed presentations of inventory uncertainty at the regional level. Problems concerning uncertainty and verification of the distributed inventory are discussed. A Monte Carlo analysis of uncertainties in the energy sector at the regional level is performed, and a number of simulations concerning the effectiveness of uncertainty reduction in some regions are carried out. Uncertainties in activity data have a considerable influence on overall inventory uncertainty, for example, the inventory uncertainty in the energy sector declines from 3.2 to 2.0% when the uncertainty of energy-related statistical data on fuels combusted in the energy industries declines from 10 to 5%. Within the energy sector, the 'energy industries' subsector has the greatest impact on inventory uncertainty. The relative uncertainty in the energy sector inventory can be reduced from 2.19 to 1.47% if the uncertainty of specific statistical data on fuel consumption decreases from 10 to 5%. The 'energy industries' subsector has the greatest influence in the Donetsk oblast. Reducing the uncertainty of statistical data on electricity generation in just three regions – the Donetsk, Dnipropetrovsk, and Luhansk oblasts – from 7.5 to 4.0% results in a decline from 2.6 to 1.6% in the uncertainty in the national energy sector inventory.

Keywords energy sector · geoinformation system · greenhouse gas · greenhouse gas inventory · multilevel model · spatial analysis · uncertainty

R. Bun (✉) · A. Bun
National University 'Lviv Polytechnics',
12 Bandera Street,
79013 Lviv, Ukraine
e-mail: rbun@org.lviv.net

M. Gusti
International Institute for Applied Systems Analysis,
2361 Laxenburg, Austria

L. Kujii · O. Tokar · Y. Tsybrivskyy
State Scientific and Research Institute of Information
Infrastructure, National Academy of Sciences of Ukraine,
P.O. Box 5446, 79031 Lviv, Ukraine

1 Introduction

The Kyoto Protocol to the United Nations Framework Convention on Climate Change (UNFCCC) defines obligations for its parties to reduce their greenhouse gas (GHG) emissions compared with those of a base year. According to the Protocol, each party must develop a national system for estimating anthropogenic emissions and sinks of GHGs. The Intergov-

🖄 Springer

ernmental Panel on Climate Change (IPCC) has developed a general methodology for estimating GHG emissions and sinks, which has been published in the Revised 1996 IPCC Guidelines (IPCC, 1997a) and corresponding software (IPCC, 1997b). A positive feature of the IPCC methodology is its universality, which allows it to be used by experts in many countries, notwithstanding these countries' different locations around the world and their different levels of economic development. This is one reason why the IPCC Guidelines have been so important during the formation of the Kyoto Protocol mechanisms.

In the future, however, this universality could slightly decrease the efficiency of GHG inventories and thus limit the use of the Kyoto mechanisms. Because of its universality, the IPCC methodology cannot consider regional disparities within countries, which could thus increase inventory uncertainty. Moreover, in most large countries, the various GHG sources and sinks are distributed nonuniformly across the territory. This is the case with Ukraine, for instance, which has an area of 603,000 square kilometers and comprises 25 administrative units (oblasts). The IPCC GHG inventory methodology gives results for entire countries and thus cannot be an effective tool for those making strategic economic and political decisions on regional development within a country.

Integrated information on the actual spatial distribution of GHG sources and sinks would aid in making well-considered economic and environmental decisions. Neighboring countries are interested in real information on ecological conditions near their borders. Geographically explicit data are needed for modeling GHG fluxes. Moreover, spatially distributed analysis of GHGs and their uncertainties can help to identify cost-effective ways of reducing uncertainty.

GHG inventories for regions within a country and the use of geographical information systems (GIS) to increase inventory quality and usability are becoming more widespread. In Portugal, for example, the national GHG inventory was carried out by region and the emissions were spatially analyzed for emission-reduction purposes (Seixas et al., 2002). There have also been efforts to disaggregate GHG emissions on a spatial grid and to produce the georeferenced maps necessary for modeling. For example, the project CARBOEUROPE-GHG (Synthesis of the European Greenhouse Gas Budget; see http://gaia.agraria.unitus.it/ceuroghg/projghg.html) disaggregates

GHG emissions to a 50×50 km grid. The project currently concentrates on the 15 original European Union (EU) member countries; however, the plan is to ultimately study the new EU countries as well, and to obtain disaggregated GHG emissions for Ukraine and Russia for full coverage of the continent. Another project is aimed at spatial disaggregation of the 1990 emissions inventory data to a 20×20 km grid for Africa south of the equator (Fleming & van der Merwe, 2000).

This article discusses bottom-up inventory analysis. We examine carbon dioxide (CO_2) emissions and their uncertainties in two dimensions – energy subsectors and spatial distribution – and determine which dimension is the most influential. A similar analysis has been performed for the Netherlands by Vreuls (2004), who considers more GHG gases and sources but omits spatial analysis. We agree with Gillenwater, Sussman, and Cohen (2007) that the uncertainty inherent in the uncertainty estimates is rather large. Nevertheless, we think that the uncertainty estimates should be used to aid policy making. Examples of practical ways of coping with the uncertainties in GHG emissions estimates when trading or comparing national GHG emissions are listed in the conclusions by Monni, Syri, Pipatti, and Savolainen (2007); valuable theoretical work is also offered by Nahorski, Horabik, and Jonas (2007).

The basic approach to carrying out a multilevel, spatially distributed inventory is considered in Section 2 of this chapter, and the geoinformation technology developed to carry out such an inventory is discussed in Section 3. Sections 4 and 5 illustrate the application of the technology for the analysis of GHG emissions in the energy sector at the regional and plot levels, respectively, while Section 6 is devoted to simulations and analysis of the uncertainties and uncertainty reduction measures. Conclusions are presented in the final section.

2 Basic Approach

The IPCC methodology (IPCC, 1997a) covers a number of human activities associated with GHG emissions and sinks – in particular, fossil fuel combustion, industry and agriculture, land-use change, and deforestation. On the basis of this methodology, we have developed a geoinformation technology that presents

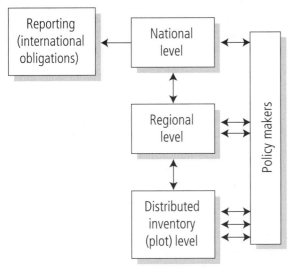

Fig. 1 Three-level structure of the inventory process

GHG inventories at three levels: the national level, the regional (oblast) level, and the plot level (Bun, 2004). Such a multilevel inventory model reflects the disparities among GHG emissions and sinks, and can be helpful for making policy decisions at the national and regional levels (Fig. 1). Information at lower inventory levels can prove extremely valuable for decision makers.

2.1 National Level

At the highest inventory level, the national level, the GHG inventory is carried out for a country as a 'point in space.' In this case, GHG inventory methods commonly used for an entire country can be utilized, following the formula

$$Y = \sum_{s=1}^{S} y_s = \sum_{s=1}^{S} \sum_{m=1}^{M_s} a_{sm} x_{sm}, \qquad (1)$$

where Y and y_s are the inventory results for the entire country and for the sth sector, respectively; S is the number of human activity sectors according to the IPCC (1997a); a_{sm} and x_{sm} are the emissions factor and data on the mth human activity in the sth sector, respectively; and M_s is the total number of human activities in the sth sector. Input data used for the inventory are taken from statistical yearbooks, research results, etc. Provided that all the necessary data are available for a country, the Revised 1996 IPCC

Guidelines permit the calculation of GHG emissions and sinks (the output data of the model). In this case, the methodology described in the Revised 1996 IPCC Guidelines and presented by expression (1) can be regarded as a mathematical model of inventory at the highest level. We have mathematical expressions mapping input data to output data, which are necessary for making an inventory for the whole country. At the highest inventory level, the input data and inventory results are 'lumped,' that is, a single value is generated for the entire country. The uncertainties are considered for the economic sectors and the country as a whole.

2.2 Regional Level

At the middle inventory level, the regional level, the inventory is carried out for each administrative region of a country. As in the previous case, the parameters of the mathematical models are lumped. Ukraine, for example, has 25 administrative regions (oblasts), some of which are the size of small countries. In principle, a methodology based on the Revised 1996 IPCC Guidelines can be applied to each region as described above, using an inventory model of the following form:

$$Y_r = \sum_{s=1}^{S} y_{rs} = \sum_{s=1}^{S} \sum_{m=1}^{M_s} a_{rsm} x_{rsm}, \qquad r = 1, \ldots, R, \quad (2)$$

where Y_r and y_{rs} are the inventory results for the rth region and its sth sector, respectively, based on the IPCC methodology (IPCC, 1997a); a_{rsm} and x_{rsm} are the emissions factor and data on the mth activity in the sth sector for the rth region, respectively; and R is the total number of regions.

Model (2) reflects regional characteristics of GHG emissions and sinks quite well, although the model parameters are lumped. Like the mathematical model for the highest level, this model has input and output data. Input data are obtained from statistical yearbooks (because most of the statistical information is published for administrative regions) and from the results of scientific research representing regional characteristics of some of the parameters used in the IPCC Guidelines. In situations where a parameter is known for the country but not for individual regions, some assumptions and additional information can be used to obtain the algorithm for determining the necessary parameters for the regions. In this case, the uncertainties are considered by economic sector and

🌍 Springer

region. We introduce additional information into the inventory (e.g., region-specific emissions factors and activity data) that decreases the overall uncertainty of the inventory at the national level; however, some regional uncertainties can be quite large.

2.3 Plot Level

At the lowest inventory level, the plot level, both input and output data are stored in a georeferenced database. This inventory level is used for plots (say, 10×10 km) covering the entire country (in the case of Ukraine, about 60,300 plots in total). For each plot, a GHG inventory is performed following the IPCC methodology (IPCC, 1997a) using a mathematical model defined according to the IPCC Guidelines.

Data on human activity in the nth plot are denoted by Δx_{nsm}, with corresponding indices. Inventory results in total and by sector for a given plot are denoted by ΔY_n and Δy_{ns}, respectively. In this case, the inventory model can be written in the following form:

$$\Delta Y_n = \sum_{s=1}^{S} \Delta y_{ns} = \sum_{s=1}^{S} \sum_{m=1}^{M_s} a_{nsm} \Delta x_{nsm},$$

$$n = 1, \ldots, N,$$

(3)

where a_{nsm} is the emissions factor for the mth activity of the sth sector in the nth plot, and N is the total number of plots. Unlike in the previous cases, in model (3), input and output data relate to individual plots; that is, they are not lumped. Some model parameters can be obtained (e.g., using a digital map and additional algorithms), and other model parameters are estimated following algorithms developed under certain assumptions.

Concerning this distributed model, in some cases the GHG emissions and sinks within a particular plot can be calculated directly using the IPCC Guidelines with corresponding emissions factors – for example, emissions from power plants, cement production plants, chemical plants, fertilized fields, etc. However, in some cases it is more efficient to distribute results obtained for a region using data on the spatial distribution of activities – for example, GHG emissions from gas flaring used for heating buildings and cooking. The GHG emissions distribution in this case correlates with population density, which is obtained from spatial analysis of a digital map (Kujii &

Oleksiv, 2003; Tsybrivskyy & Klym, 2003). In the worst case, if one cannot derive detailed data on GHG emissions caused by specific human activities within a region, the total emissions quantity for all plots within the region can be distributed uniformly. GHG inventories at the plot level include more information than those at the national and regional levels (e.g., location of stationary emissions sources, spatial distribution of sources and sinks, usage of plant-specific emissions factors and activity data, etc.) and thus decrease the overall uncertainty.

In summary, in the modeling approach presented here the distributed inventory is carried out for a selected class of objects (regions, districts, or plots). Information obtained from layers of a digital map and statistical data for regions and districts are used as input data. From this distributed inventory, new layers of a digital map are formed corresponding to the economic sectors of the IPCC methodology. Summing inventory results for all plots within Ukraine produces a general inventory for the entire country (Bun et al., 2002, 2003).

The technology used is based on a GIS, the IPCC methodology, and special software. The use of digital maps and the geoinformation approaches makes possible a distributed inventory of the territory, while the use of the IPCC methodology and software means the inventory results are compatible and comparable with those of traditional approaches. Moreover, the use of region-specific emissions factors and activity data increases the quality of the GHG inventory (Bun, 2004).

3 A Geoinformation Technology for Distributed GHG Inventories

The geoinformation technology presented in the previous section, which combines georeferenced databases, geoinformation systems, and the IPCC methodology, is illustrated in more detail in Fig. 2, which shows the corresponding layers of a digital map. Here, information from these layers together with statistical data and data from scientific research serve as input data. The databases (i.e., the new layers of the digital map) corresponding to the economic sectors of the IPCC methodology (energy, industrial processes, etc.) are created using this input information.

We perform the inventory using the IPCC methodology for all plots within a given country. In this

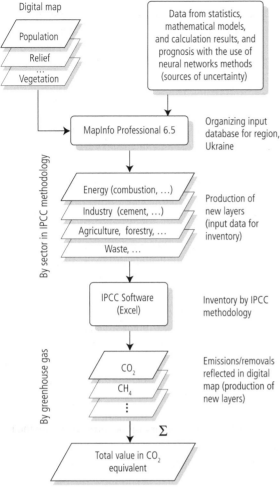

Fig. 2 Geoinformation approach to GHG inventory

classifiers, and is realized in the MapInfo system format. The following segments were used to create an inventory of GHG emissions: settlements (inhabited localities and their population), forested lands, hydrology, oblast boundaries, vegetation, and soil.

Statistical data published by the State Committee of Statistics of Ukraine in a number of statistical collections (e.g., Ukrstat, 2001) are another major source of input information for the GHG inventory. The statistical data are issued for many economic sectors, and the information is aggregated for oblasts. Regional statistical collections also exist. Thus it is an easy step from GHG inventories at the national level to those at the oblast level. On the basis of the statistical collections, one can obtain the input data necessary to complete the input worksheets of the IPCC methodology by economic sector.

The proposed geoinformation system consists of two basic modules: GHGinvent and GHGmap (Bun & Oleksiv, 2003). GHGinvent is a programming module that performs a GHG inventory according to a user-defined inventory model (i.e., at the selected level). The main function of GHGinvent is to input data into the corresponding Excel tables in the IPCC methodology (IPCC, 1997b). This module forms initial GHG inventory tables using the results of the IPCC methodology according to the model used.

The basic functions of the GHGmap module are to organize queries into inventory tables and to form new geoinformation layers with the inventory results, which are then reflected in the digital map of Ukraine. The inventory tables organized by GHGinvent, together with the topographical information of the digital map of Ukraine, serve as input data for GHGmap. The proposed geoinformation technology is quite complex with respect to software implementation because a number of different kinds of software components have to interact correctly if the entire information system is to perform as it should. The software includes databases of input information filled in by the operator (Bun & Oleksiv, 2003); Excel tables of the IPCC methodology filled in by the program according to the inventory model used; and database tables that are compatible with MapInfo for inventory results reflected in the digital map of Ukraine. Below, a number of results of GHG inventories at the regional and plot levels as well as a spatial uncertainty analysis are presented to illustrate possible ways of using the technology.

way, we form the new layers of a digital map corresponding to the results of the GHG inventory of emissions of carbon dioxide (CO_2), methane (CH_4), etc. Finally, we obtain the layer of the digital map corresponding to the total GHG emissions in CO_2 equivalent terms. Thus, in the proposed approach to creating a distributed inventory, the results are produced in the form of layers of a digital map of Ukraine. Lower-level inventory results include information about specific levels of GHG emissions and sinks per unit area within the country.

The digital map of Ukraine produced by Intelligence Systems GEO Ltd. (ISGEO: http://www.isgeo.kiev.ua) was chosen for use in the proposed geoinformation technology. The map is a spatial database at a 1:500,000 scale. The database is organized in the form of separate tables containing cartographic objects and

4 Inventory at the Regional Level: Energy Sector

Let us consider GHG inventories at the regional level, using as an example the energy sector, which accounts for about 95% of total GHG emissions in Ukraine. The primary sources of GHGs in the energy sector are fuel production, fuel transportation, and fuel combustion (Kujii, 2003). The Revised 1996 IPCC Guidelines (IPCC, 1997a) for the energy sector cover six GHGs: NO_2, CH_4, nitrous oxide (N_2O), carbon monoxide (CO), nitrogen oxides (NO_x), non-methane volatile organic compounds (NMVOCs), and sulfur dioxide (SO_2). CO_2, CH_4, and NO_x emissions are the largest. Below we consider a sectoral approach; that is, we account for the carbon in fuels supplied to the economic sectors (IPCC, 1997a).

CO_2 emissions resulting from fuels combusted in the energy industries (i.e., fuel-extraction or energy-producing industries; for details, see IPCC, 1997a, Vol. 1) determine the sector emissions (Kujii, 2003). With integrated statistical data on production, export, import, and consumption of fuel and energy resources, one can use the IPCC Guidelines (IPCC, 1997a) to estimate the carbon mass (in gigagrams [Gg]) in the utilized fuel:

$$m_c = mTk_c, \qquad (4)$$

where m is the mass of combusted fuel (in Gg), T is the fuel calorific value (in terajoules [TJ] per Gg), and k_c is the carbon emissions factor (in tons of carbon per TJ). The fraction of nonoxidized carbon (f_c) should also be accounted for.

Models that create an inventory of GHG emissions from fuel combustion by economic sector (energy industries; manufacturing industries and construction; international marine and air transport; the commercial/institutional and residential sectors; agriculture/forestry, etc.) provide more useful information than more aggregated models, since different coefficients – T, k_c, and f_c – are applied to different economic sectors.

In many countries, natural, historical, and other factors have led to the nonuniform distribution of GHG emissions from the energy sector. This is true of Ukraine, which has developed industrial regions with high consumption of fuel and energy resources, as well as of regions without heavy industry. The technology for creating spatially distributed inventories is useful for presenting these differences in GHG emissions at the regional level. Results of an inventory of CO_2 emissions caused by fuels combusted in the energy industries at the regional level are presented in Fig. 3. The data relate to the economic activity of the regions of Ukraine in 2000. The emissions are very irregularly distributed; thus, for convenient presentation of the results, a square root function of the data is used (the

Fig. 3 Variation of GHG emissions resulting from fuels combusted in the energy industries among the regions of Ukraine

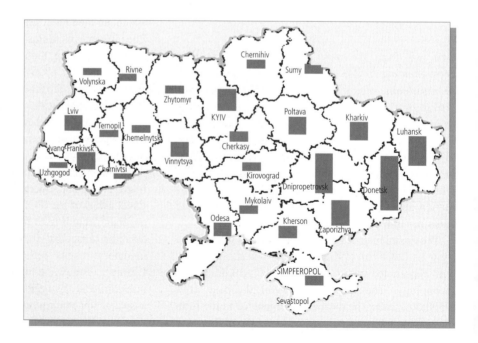

column height is proportional to the square root of the emissions value).

The Donetsk oblast has the highest CO_2 emissions in Ukraine (26.89% of total CO_2 emissions). Half the total emissions (51.84%) are contributed by three oblasts: Donetsk, Luhansk, and Dnipropetrovsk. Most of the CO_2 emissions occur in the processes of the energy industries. The difference between the results of CO_2 emissions obtained using the reference (accounting for the carbon in fuels supplied to the entire economy) and sectoral approaches does not exceed 10%. The discrepancy between the emissions estimates using the two different approaches can be explained by the fact that statistical data for sectors are set equal to zero in cases where their values are below the lowest-order number in the corresponding statistical table. This phenomenon occurs where fuel-energy resources are presented by oblast or economic activity. Therefore, total emissions do not always equal the sum of the individual components. The small discrepancy between the calculated emissions values in the oblasts allows the user to draw conclusions as to the consistency of the statistical data on fuels combusted in the energy industries of Ukraine (Bun, 2004).

In 2000, for all GHGs, the highest emissions were observed in the Donetsk (109,669 Gg of CO_2 equivalent), Dnipropetrovsk (56,607 Gg of CO_2 equivalent),

and Luhansk (41,964 Gg of CO_2 equivalent) oblasts. CO_2 sink values exceeded emissions values in a number of oblasts, particularly in the Volynska (2,348 Gg of CO_2 equivalent), Zakarpatska (Uzhgorod) (4,821 Gg of CO_2 equivalent), Rivne (1,828 Gg of CO_2 equivalent), Chernivtsi (92 Gg of CO_2 equivalent), and Chernihiv (1,897 Gg of CO_2 equivalent) oblasts. The emissions levels are determined mainly by the energy sector; absorption levels, by the land use change and forestry sector (Bun, 2004).

Within the energy sector, the lowest CO_2 emissions are from natural gas combustion, as it has the lowest emissions factor (approximately half that of coal) (IPCC, 1997a). Thus, the shift from coal to natural gas and black oil in combined heat and power (CHP) plants could solve the GHG problem for Ukraine's energy sector. Taking into account the significant contribution of CHP plants to the total GHG budget of Ukraine, plans for the development of domestic sources of electricity and heat supply must be revised. Increasing the efficiency of power equipment will help solve this problem (Bun, 2004).

5 Spatial Analysis of GHG Emissions

In carrying out the distributed inventory, each plot of Ukrainian territory is analyzed in turn. If the border

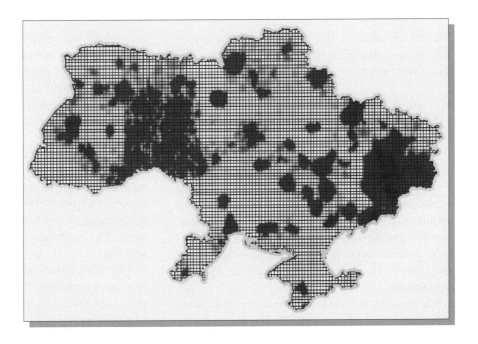

Fig. 4 Presentation of CO_2 emissions resulting from combustion of coal in the public sector at the plot level (distributed inventory); *darker areas* indicate higher emissions levels

between two or more administrative units lies within a plot, the emissions and sinks are assigned in proportion to each unit's contribution.

CO_2 emissions resulting from the combustion of coal in the public sector at the plot level (10×10 km plots, distributed inventory results) are shown in Fig. 4. The figure only gives qualitative information on the territorial distribution of the emissions; however, the digital layer comprises the data in each plot that can be used for analysis. This type of digital layer can be made for each GHG and for each kind of human activity considered in the IPCC methodology (IPCC, 1997a).

Moreover, the geoinformation technology allows the user to make projections of GHG emissions and sinks following different scenarios of economic development. In the most favorable scenario, the GHG emissions reach their 1990 level in 2011–2012 (Bun, 2004). Emissions reach this level in 2013–2014 in the favorable scenario and in 2020 in the unfavorable scenario. As the unfavorable scenario corresponds to slow changes in the economy, it is the most likely scenario.

The territorial approach to constructing CO_2 inventories takes into account regional differences in economic activities within Ukraine. The multilevel inventory is aimed at obtaining quantitative estimates for separate regions of the country. Estimates of distributed GHG emissions (on a territorial basis) from the energy sector can help to accelerate the implementation of actions to reduce emissions – for example, means for GHG utilization, the capture of CO_2 from exhaust, the creation of favorable conditions for carbon absorption by forests, etc.

6 Results of Spatial Inventory and Uncertainty Reduction

In carrying out national inventories and in trading emissions permits, one must be sure that inventory results are of 'good quality' (i.e., that the uncertainties are small). All data used in the inventory (emissions factors, calorific values, statistical activity data, etc.) have some uncertainty that can significantly slow the process of implementing the Kyoto Protocol mechanisms.

Uncertainty in GHG inventories is the value indicating the lack of certainty in the cadastre

components resulting from such arbitrary random factors as uncertainty of emissions sources, lack of transparency in the inventory process, etc. (IPCC, 2000). Most often, relative uncertainty is characterized as a 95% confidence interval, meaning that the probability that the value of a real parameter falls within the interval is 95%. Relative uncertainty is 'measured' in percent as the ratio of the confidence interval value to the mean parameter value. If every value used in the GHG inventory has some uncertainty, then the inventory process according to the IPCC methodology (IPCC, 1997a), which utilizes multiplication and summing, leads to an 'uncertainty combination' in compliance with the following formulas (IPCC, 2000):

$$U_{total} = \frac{\sqrt{(U_1 \cdot x_1)^2 + (U_2 \cdot x_2)^2 + \ldots + (U_k \cdot x_k)^2}}{x_1 + x_2 + \ldots + x_k} \tag{5}$$

for the uncertainty of the sum of values $x_1 + x_2 + \ldots + x_k$, and

$$U_{total} = \sqrt{U_1^2 + U_2^2 + \ldots + U_k^2} \tag{6}$$

for the uncertainty of the product of the values. The resulting uncertainty is given as a percentage; x_i and U_i are the uncertain value and its relative uncertainty, respectively (in percent).

The formulas presented above for the uncertainty combination relate to the case of a normal distribution of random uncorrelated values. The Monte Carlo method is more general and consists of choosing random values of emissions factors and activity data from their individual probability distributions and calculating corresponding emissions (IPCC, 2000). This procedure is repeated many times, and the results of all iterations form the probability distribution of emissions. A Monte Carlo analysis can be conducted for every emissions source for economic sectors, national regions, or the entire cadastre. The Monte Carlo method allows the user to work with probability distributions of any form and to account for correlations. The experiment results presented below were obtained using this method.

The geoinformation technology developed for creating a multilevel inventory allows the user to carry out experiments on the uncertainties in national GHG inventories in, for example, the energy sector

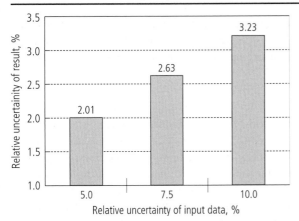

Fig. 5 Influence of activity data uncertainty on the uncertainty of national inventory in the energy sector

and to determine the dependence of the uncertainties on inventory components. Using this feature highlights ways of reducing the uncertainties. A number of such experiments are discussed below.

Experiment 1 A report by the IPCC (2000) provides uncertainty intervals for statistical data for countries such as Ukraine. For data on fuels combusted in the energy industries (which largely determines GHG emissions in Ukraine), the interval provided by the IPCC is 5–10% – a more exact uncertainty value should be found by national experts. Using these recommended uncertainty intervals, we carried out an experiment on the influence of activity data uncertainty on the uncertainty of the national inventory in the energy sector. Figure 5 shows the results based on economic activity in the regions of Ukraine in 2000 for three uncertainty values from the uncertainty interval recommended by the IPCC (energy industries) – the lowest (5.0%), middle (7.5%), and highest (10.0%) interval values. In all experiments reported here, the uncertainties in other sectors were assumed to be the mean of the intervals recommended by the IPCC (2000) (see specifications in Experiment 2).

It was assumed that the statistical data on economic activity in the energy sector were of a normal probability distribution and that their uncertainty, characterized by a 95% confidence interval, was similar in all regions. Data on the calorific value of the fuel were assumed to have a normal probability distribution and 5% uncertainty for the 95% confidence interval. The other data used in the inventory were assumed to be known exactly. The calculations of national emissions in the energy sector were carried

out many times for different randomly chosen inventory parameters for the regions of Ukraine. The probability distribution of the parameters for the national inventory in the energy sector is determined from the calculated results. The results show that decreasing the uncertainties in national statistics is valuable for implementing the Kyoto Protocol mechanisms. The uncertainty of national energy sector inventory data decreases from 3.2% (for higher uncertainty of statistical data – the 10% interval value) to 2.0% (for lower uncertainty of statistical data – the 5% interval value). This leads to a change in the confidence interval of 8.4 Gg of CO_2.

Experiment 2 Calculation results demonstrating the dependence of national energy sector inventory uncertainty (in percent) on the uncertainty in each subsector are shown in Fig. 6. The estimation is carried out using data on economic activity in Ukraine in 2000 for minimum and maximum uncertainties of the statistical data for each subsector as follows: (1) fuels combusted for energy production (5–10%); (2) manufacturing industries and construction (5–10%); (3) transport (5–10%); (4) commercial/institutional, and residential sectors (15–20%); (5) agriculture/forestry (5–10%); and (6) other (15–20%). The minimum and maximum uncertainties are taken from the uncertainty intervals recommended by the IPCC (2000). When the simulation was carried out for one of the subsectors, the uncertainties for the other subsectors were chosen to be the mean of the

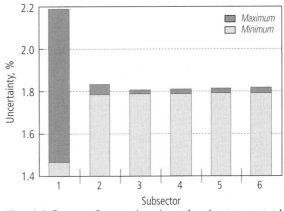

Fig. 6 Influence of uncertainty in each subsector on total inventory uncertainty in the energy sector: (*from left*) *1* = energy industries; *2* = manufacturing industries and construction; *3* = transport; *4* = commercial/institutional and residential sectors; *5* = agriculture/forestry; *6* = other

recommended intervals. The other parameters are defined as in Experiment 1.

The results demonstrate that considerable emissions result from fuels combusted in the energy industries and that decreasing the uncertainty in this subsector is an urgent problem. Specifically, the relative uncertainty in the national inventory in the energy sector can be reduced from 2.19 to 1.47% (when absolute uncertainty equals 2.5 Gg of CO_2).

Experiment 3 Continuing from Experiment 2, an analysis was carried out on how energy sector inventory uncertainty in each region contributes to the total inventory uncertainty (in absolute values). The results are presented in Table 1. The greatest influence appears in the 'energy industries' subsector in the Donetsk oblast, where the absolute uncertainty is 5,081 Gg of CO_2 (2.23% uncertainty relative to the total

CO_2 emissions in this subsector; that is, 227,819.40 Gg of CO_2), followed by the Dnipropetrovsk and Luhansk oblasts, with uncertainties in these subsectors equaling 2,066 Gg of CO_2 (0.91%) and 1,262 Gg of CO_2 (0.55%), respectively.

Experiment 4 The improvement of statistics (i.e., decreasing the uncertainty in statistical data) requires considerable investments, such as the installation of additional equipment, the implementation of organizational and administrative measures for a more accurate and complete record of all economic spheres, and additional research for a better understanding of emissions processes. Thus, those administrative regions that have the most influence on energy sector emissions should be identified, and investments for decreasing the uncertainty in statistical data should be increased only in these regions. As these regions have

Table 1 Absolute uncertainty of the inventory in the energy sector in Ukrainian regions

N	Region	Absolute uncertainty (Gg of CO_2)					
		Energy industries	Manufacturing industries and construction	Commercial/institutional sectors	Agriculture	Forestry	Other
1	Cherkasy	118.39	18.45	49.47	16.56	47.89	119.14
2	Chernihiv	88.12	2.95	14.16	18.20	36.27	116.74
3	Chernivtsi	21.77	2.09	7.68	8.49	8.71	71.98
4	Crimea	133.64	12.45	34.93	7.06	37.82	122.63
5	Dnipropetrovsk	2066.48	1056.68	44.10	79.71	57.52	540.24
6	Donetsk	5081.51	904.08	56.45	180.32	55.11	468.26
7	Ivano-Frankivsk	447.06	36.14	121.85	11.41	7.16	191.49
8	Kharkiv	563.32	49.57	31.80	38.32	61.70	286.62
9	Kherson	384.43	7.19	11.71	11.64	40.49	87.47
10	Khmelnytsk	61.31	26.81	16.14	26.01	51.64	143.54
11	Kirovograd	106.35	3.01	8.72	9.33	43.83	89.30
12	Kyiv	924.67	30.81	58.87	19.19	70.06	397.13
13	Luhansk	1261.77	381.40	41.55	49.85	33.27	372.73
14	Lviv	314.83	32.71	22.91	56.21	15.65	412.43
15	Mykolaiv	115.99	28.61	14.75	0.38	41.29	114.79
16	Odesa	361.43	9.33	46.32	17.12	45.61	143.52
17	Poltava	673.93	33.67	83.15	17.07	57.21	299.36
18	Rivne	55.01	29.00	14.24	12.87	19.07	81.18
19	Sumy	116.07	21.28	62.12	12.98	39.35	140.12
20	Ternopil	49.69	5.74	7.88	6.40	26.84	118.82
21	Zakarpatska (Uzhgorod)	23.08	1.96	10.28	1.16	2.41	104.45
22	Vinnytsya	385.56	7.09	17.25	38.44	64.60	240.66
23	Volynska	47.69	2.80	7.57	8.32	17.46	87.86
24	Zaporizhya	1091.18	337.79	28.26	25.64	47.19	131.52
25	Zhytomyr	54.59	14.95	18.25	27.39	34.63	114.49

the most influence on emissions, decreasing uncertainty here will lead to a decrease of uncertainty in the national inventory.

As shown in Fig. 3, the CO_2 emissions resulting from fuels combusted in the energy industries are the highest in the Donetsk, Dnipropetrovsk, and Luhansk oblasts. According to the IPCC recommendations (IPCC, 2000), uncertainty in 'better-developed' statistics is within a 3–5% interval. Thus, the influence of investments was studied only with respect to improving statistics relative to CO_2 emissions from fuels combusted in the energy industries and only in those regions where decreased inventory uncertainty would reduce the national energy sector GHG inventory uncertainty (Fig. 7).

The uncertainty values shown in Fig. 7 relate to economic activity in Ukraine in 2000. Column 1 illustrates the initial uncertainty of the national inventory in the energy sector if the statistical data in all regions have a mean uncertainty from the IPCC (2000) interval for 'poorly developed' statistical systems (7.5% for CO_2 emissions from fuels combusted in the energy industries; other parameters are defined as in Experiment 1). Column 2 corresponds to the case where the uncertainties of all data in all regions remain unchanged, except for the uncertainties of statistical data in the Donetsk oblast, which decrease to 4%, the mean value from the uncertainty interval recommended for countries with a 'well-developed' statistical system (IPCC, 2000). Column 3 corresponds to the case where the uncertainty is decreased to 4% in two regions: the Donetsk and Dnipropetrovsk oblasts. Column 4 relates to the case where the uncertainty is decreased in the third region (Luhansk oblast) as well. The decline of uncertainty in the national inventory from 2.6 to 1.9%

(a considerable decrease of uncertainty in absolute values presented in Fig. 7 is achieved just by decreasing uncertainty in only one activity type in three regions.

7 Conclusions

The IPCC methodology (IPCC, 1997a) provides inventory methods for entire countries. From the international viewpoint, such inventories make sense. However, every government should also have tools for exploring the real situation at the regional level. The proposed geoinformation technology for creating a multilevel distributed inventory allows GHG emissions cadastres to be created at both the regional and the plot level (covering the entire country). Integrated information on the actual spatial distribution of GHG sources and sinks would be quite useful for decision makers. Such information and corresponding visualization tools could serve as an effective instrument in economic and environmental decision making. Features of the proposed geoinformation technology include the following:

- The technology reflects the real state of GHG emissions and sinks at the regional level.
- It is based on the use of digital maps and the IPCC methodology, combining inventory transparency and ease of documentation.
- It allows the effective utilization of remote-sensing data, neural network technologies, and approaches to estimating and projecting a number of parameters of distributed models of processes of GHG emissions and sinks at the regional level.
- It is effective for large countries with nonuniformly distributed GHG sources and sinks, and thus is a good instrument for regional management decision making and for carrying out projections in accordance with development strategies, including sustainable development strategies.

The decrease of uncertainties in national statistics has a considerable influence on inventory uncertainty. For example, the inventory uncertainty in the energy sector declines from 3.2 to 2.0% when the uncertainty of energy-related statistical data on fuels combusted in the energy industries declines from 10 to 5%.

Within the energy sector, the 'energy industries' subsector has the greatest impact on inventory uncertainty. The relative uncertainty in the national inventory in the energy sector could be decreased from 2.19

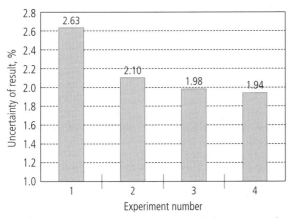

Fig. 7 An example of uncertainty decrease in inventory results

Springer

to 1.47% if the specific statistical data uncertainty on combusted fuels were to decrease from 10 to 5%.

This subsector has the largest influence in the Donetsk oblast, where the absolute uncertainty is 5,081 Gg of CO_2 (2.23% uncertainty relative to the total CO_2 emissions in this subsector). Second and third are the Dnipropetrovsk and Luhansk oblasts, with uncertainties in this subsector of 2,066 Gg of CO_2 (0.91%) and 1,262 Gg of CO_2 (0.55%), respectively.

Improving the statistical system, especially for the 'energy industries' subsector in three regions of Ukraine (Donetsk, Dnipropetrovsk, and Luhansk oblasts) in order to decrease the uncertainty of the statistical data from 7.5% (for a 'poorly developed' statistical system) to 4% (for a 'better-developed' statistical system) will result in a decline in the uncertainty in the national energy sector inventory from 2.6 to 1.6%.

The geoinformation technology for creating distributed inventories proposed here enables the most essential sources of uncertainty to be defined (kinds of activity and regional locations) and makes possible the more effective utilization of investments to reduce uncertainty in these locales and in these kinds of activity. Certainly, for the geoinformation technology for spatially distributed inventories, some new problems arise concerning uncertainty and verification, but this technology allows for qualitative and quantitative 'distributed' presentation of the uncertainty problem at the regional level.

References

Bun, R. (Ed.) (2004). *Information technologies for greenhouse gas inventory and prognosis of carbon budget of Ukraine.* Lviv, Ukraine: UAP.

Bun, R., Gusti, M., Dachuk, V., Oleksiv, B., & Tsybrivskyy, Y. (2003). Specialized computer system for multilevel inventory of greenhouse gases. *Herald of Technological University of Podillia, 3,* 77–81.

Bun, R., & Oleksiv, B. (2003). Specialized database for information technologies of greenhouse gases inventory. *Information Technologies and Systems, 1–2,* 195–201.

Bun, R., Oleksiv, B., Klym, Z., Kujii, L., & Tsybrivskyy, Y. (2002). Geoinformation systems as a tool for carbon cycle monitoring and greenhouse gas inventory in the Western Region of Ukraine. In *Proceedings of the International Conference 'Mountains and People,'* Vol. 2, *'In the Context of Sustainable Development,* Rakhiv, Ukraine, October 2002, pp. 21–25.

Fleming, G., & van der Merwe, M. (2000). Spatial disaggregation of greenhouse gas emission inventory data for Africa South of the equator. *CSIR,* Pretoria, South Africa. Available at http://gis.esri.com/library/userconf/proc00/professional/papers/PAP896/p896.htm.

Gillenwater, M., Sussman, F., & Cohen, J. (2007). Practical policy applications of uncertainty analysis for national greenhouse gas inventories. *Water, Air, & Soil Pollution: Focus* (in press). doi:10.1007/s11267-006-9118-2

IPCC (1997a). *Revised 1996 IPCC Guidelines for National Greenhouse Gas Inventories: Reporting Instructions, the Workbook, Reference Manual,* Vol. 1–3, IPCC/OECD/IEA, Intergovernmental Panel on Climate Change (IPCC) Working Group I (WG I) Technical Support Unit, Bracknell, UK. Available at http://www.ipcc-nggip.iges.or.jp/public/gl/invs4.htm.

IPCC (1997b). IPCC *Greenhouse Gas Inventory Software for the Workbook,* IPCC. Intergovernmental Panel on Climate Change (IPCC), Geneva, Switzerland. Available at http://www.ipcc-nggip.iges.or.jp/public/gl/software.htm.

IPCC (2000). *Good Practice Guidance and Uncertainty Management in National Greenhouse Gas Inventories,* Intergovernmental Panel on Climate Change (IPCC) National Inventories Programmes, Technical Support Unit, Institute for Global Environmental Strategies, Hayama, Kanagawa, Japan. Available at http://www.ipcc-nggip.iges.or.jp/public/gp/english/.

Kujii, L. (2003). Models for greenhouse gas inventories in the energy sector of Ukraine. *Information Technologies and Systems, 1–2,* 202–210.

Kujii, L., & Oleksiv, B. (2003). Methods and means for realization of geoinformation system of greenhouse gases inventory. *Scientific Papers of the Institute for Modelling Problems in Energetics, 19,* 182–192.

Monni, S., Syri, S., Pipatti, R., & Savolainen, I. (2007). Extension of EU emissions trading scheme to other sectors and gases: Consequences for uncertainty of total tradable amount. *Water, Air, & Soil Pollution: Focus* (in press). doi:10.1007/s11267-006-9111-9

Nahorski, Z., Horabik, J., & Jonas, M. (2007). Compliance and emissions trading under the Kyoto Protocol: Rules for uncertain inventories. *Water, Air, & Soil Pollution: Focus* (in press). doi:10.1007/s11267-006-9112-8

Seixas, J., Martinho, S., Gois, V., Ferreira, F., Moura, F., Furtado, C., et al. (2002). 'Greenhouse Gases in Portugal: Emissions/Economics and Reduction Measures,' *11th International Emission Inventory Conference: 'Emission Inventories – Partnering for the Future,'* held in Atlanta, GA, USA.

Tsybrivskyy, Y., & Klym, Z. (2003). Information technologies for greenhouse gas inventories: Pilot analysis of emission sources in industry and agriculture of Ukraine. *Information Technologies and Systems, 1–2,* 183–194.

Ukrstat (2001). *Statistical Yearbook of Ukraine for 2000,* State Committee of Statistics of Ukraine (Ukrstat), Tehnika, Kyiv, Ukraine.

Vreuls, H. (2004). Uncertainty analysis of Dutch greenhouse gas emission data, a first qualitative and quantitative (TIER 2) analysis. In *Proceedings of the International Workshop on Uncertainty in Greenhouse Gas Inventories: Verification, Compliance and Trading,* held 24–25 September in Warsaw, Poland, pp. 34–44. Available at http://www.ibspan.waw.pl/GHGUncert2004/papers.

Water Air Soil Pollut: Focus (2007) 7:495–511
DOI 10.1007/s11267-006-9113-7

Prior to Economic Treatment of Emissions and Their Uncertainties Under the Kyoto Protocol: Scientific Uncertainties That Must Be Kept in Mind

M. Jonas · S. Nilsson

Received: 28 May 2006 / Accepted: 20 December 2006 / Published online: 7 February 2007
© Springer Science + Business Media B.V. 2007

Abstract In a step-by-step exercise – beginning at full greenhouse gas accounting (FGA) and ending with the temporal detection of emission changes – we specify the relevant physical scientific constraints on carrying out temporal signal detection under the Kyoto Protocol and identify a number of scientific uncertainties that economic experts must consider before dealing with the economic aspects of emissions and their uncertainties under the Protocol. In addition, we answer one of the crucial questions that economic experts might pose: how credible in scientific terms are tradable emissions permits? Our exercise is meant to provide a preliminary basis for economic experts to carry out useful emissions trading assessments and specify the validity of their assessments from the scientific point of view, that is, in the general context of a FGA-uncertainty-verification framework. Such a basis is currently missing.

Keywords Kyoto protocol · full greenhouse gas accounting · uncertainty · verification · emissions · emission changes · signal detection · emission limitation or reduction commitments · risk of not meeting commitments

M. Jonas (✉) · S. Nilsson
International Institute for Applied Systems Analysis,
2361 Laxenburg, Austria
e-mail: jonas@iiasa.ac.at

1 Introduction

Full carbon accounting (FCA) or full greenhouse gas accounting (FGA),[1] uncertainty, and verification, in connection with the detection of greenhouse gas (GHG) net flux changes (also termed net flux signals), are crucial issues for the functioning of the Kyoto Protocol (Grassl et al., 2003; Nilsson et al., 2000; Nilsson, Jonas, Obersteiner, & Victor, 2001; Nilsson, Jonas, & Obersteiner, 2002; Nilsson et al., 2007; Schulze, Valentini, & Sanz, 2002; Steffen et al., 1998; Valentini et al., 2000). However, we must observe that these issues are not being concomitantly and rigorously discussed in a holistic context among or between physical scientists and experts from other disciplines (e.g., economics). Physical scientists do not

---

[1]FCA refers to a full carbon budget that encompasses and integrates all carbon-related components of all terrestrial ecosystems and is applied continuously in time. The components are typically described by adopting the concept of pools and fluxes to capture their functioning. The reservoirs can be natural or human-impacted and internally or externally linked by the exchange of carbon as well as other matter and energy. Net biome production (NBP) is the critical parameter to consider for long-term (decadal) carbon storage. NBP is only a small fraction of the initial uptake of CO_2 from the atmosphere and can be positive or negative; at equilibrium it is zero (Steffen et al., 1998, p. 1393; Jonas et al., 1999, p. 9; Nilsson et al., 2000, pp. 2, 6–7; Shvidenko & Nilsson, 2003, Section 2). FGA simply extends the definition of FCA to include other relevant GHGs (Nilsson et al., 2007, Section 1). However, a clear agreement on which gases are included is still outstanding.

scrutinize, in a holistic context, the basis that has been set by the political negotiators of the Protocol, nor do they specify the scientific constraints under which the Protocol will operate. There are many consequences of this. To safeguard their carbon trading assessments from an uncertainty-risk point of view, experts from financial institutions might, for example, ask questions that physical scientists cannot answer, such as: how credible in scientific terms are tradable emissions permits? Economics experts typically carry out assessments that are not integrated within a proper physical scientific FGA framework (i.e., they cannot properly specify the validity of their assessments from a physical scientific [verification-related] point of view). Moreover, scientists, for their part, fail to assemble crucial knowledge that will prove useful in improving the Protocol prior to and for its follow-up commitment periods. In this context, we refer to recently completed collaborative work on the preparatory detection of uncertain GHG emission signals under the Kyoto Protocol (Jonas et al., 2004a) that should have been applied before/during negotiation of the Kyoto Protocol and that addresses the question: how well do we need to know what net emissions are if we want to detect a specified emissions signal at a given point in time?

This work advances the emission reporting of Annex I countries under the Protocol, as it takes uncertainty and its consequences into consideration, that is, 1) the risk that a country's true emissions in the commitment year/period are above its true emissions limitation or reduction commitment (i.e., the risk that the country will not meet its commitment); and 2) the detectability of the country's target. The authors' approach can be applied to any net emitter, and in our follow-up work, (Jonas et al. 2004b and 2004c), we demonstrate how evaluation, in terms of risk and detectability, of GHG emission signals can become standard practice. These two qualifiers can be determined and could indeed be accounted for in pricing GHG emissions permits.

We use our preparatory signal detection work as an example in an exercise that identifies step by step beginning at FGA and ending with signal detection the relevant physical scientific constraints and choices that are involved in applying signal detection within an FGA-uncertainty-verification framework. In other words, our signal-detection results can be properly evaluated against a solid physical scientific back-ground. Our primary intention in this exercise is not to undermine the Protocol, which is not placed within such a framework and has also not been subject to preparatory signal detection, but to compensate for the lack of lucidity in the thinking behind the Kyoto Protocol and the conditions under which it will operate, including the consequences that it will have.

Moreover, our signal-detection results are of practical use. Emission signals that are assessable in terms of detectability or statistical significance have a direct bearing on how carbon permits are evaluated economically. Thus, our second intention is to use our work to build a bridge from the physical sciences to economics, that is, to offer properly specified, physical–scientific uncertainty and risk-related information that can be used by economic experts when they are working out the details of emissions trading.

Our paper is structured as follows: in Section 2 we set the stage for working within a consistent FGA-uncertainty-verification framework. In Section 3 we expose the reader to the verification of emissions in the context of bottom–up and top–down accounting. In Section 4 we explain how we merge bottom–up/top–down verification of emissions and temporal signal detection. In Section 5 we present the quantitative results of two fundamentally different preparatory signal-detection techniques and illustrate the far-reaching consequences of dealing with uncertain emission signals. Finally, in Section 6, we summarize the lessons drawn from our step-by-step analysis and establish the background against which we evaluate our signal-detection results.

Our paper is strongly guided by science–theoretical considerations and attempts to present a number of issues in a holistic context, something that has not, to our knowledge, been done elsewhere. While longer discussions of each of the issues is required, we have chosen to keep Sections 2 to 5 short to facilitate reading. However, we insert cross-references, which direct the reader to additional background information where the issues are discussed in greater depth.

2 Setting the Stage for Working within a Consistent FGA-Uncertainty-Verification Framework

In this section we develop an understanding of plausibility, validation, and verification based on our favorite way of categorizing uncertainty (Section 2.1);

we explain accounting versus diagnostic and prognostic modeling in terms of uncertainty (Section 2.2); and we specify the concept as well as the classes that we apply in order to grasp uncertainty quantitatively (Sections 2.3 and 2.4, respectively).

2.1 A Brief Science–Theoretical Discourse: Plausibility, Validation, and Verification

To illustrate the origin of uncertainties, we follow Moss and Schneider (2000; see also Giles, 2002), who categorized uncertainties and espoused the use of a straightforward concept within the Intergovernmental Panel on Climate Change (IPCC). The authors' concept reveals the advantage of fundamental structure: it considers four main categories – corresponding to confidence in the theory, the observations, the model results, and the consensus (understood as soft knowledge) within a field – to which we attach scientific quality labels to indicate whether plausibility, validation, or verification (in ascending order of scientific strictness) can be achieved (see Fig. 1; for comparison see also Vreuls, 2004, Fig. 1; and Gillenwater, Sussman, & Cohen, 2007, Section 2.2). These are specified – in line with science theory (e.g., Lauth & Sareiter, 2002) – according to the definitions used in *Merriam-Webster's Collegiate Dictionary* (Merriam-Webster, 1973 and 1997):

> *Plausibility* (from *plausibilis* = worthy of applause) → plausible: reasonable; appearing worthy of belief <the argument was both powerful and ~>.
> *Validation* (from *validus* = strong) → valid: well grounded or justifiable: being at once relevant and meaningful <a ~ theory>; logically correct (i.e., having a conclusion correctly derived from premises) <a ~ argument>.
> *Verification* (from *verus* = true) → verify: to establish the truth, accuracy, or reality.[2]

In accordance with these definitions, only observations (measurements) that are uncertain per se can be verified; none of the other categories can be verified. Theories and diagnostic models can only be validated

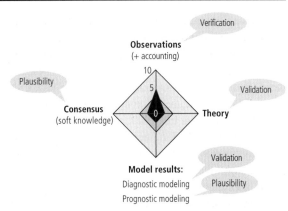

Fig. 1 Scientific quality attached to the four-axis concept of Moss and Schneider (2000, Fig. 5; see also Giles, 2002, p. 477). The figure, designed to trace where uncertainty comes from, is modified to show which scientific quality in terms of plausibility, validation, and verification can be achieved. The authors use a scale of 1–10 to reflect experts' assessments of the amount/quality of, for instance, theory and observations, to support their findings. See text for explanations

or, alternatively, falsified (which is a controversial issue in its own right). Both consensus and prognostic modeling also give rise to uncertainty. However, these two categories can, at best, be judged only as plausible; they can be neither validated nor verified.

Considering that, in the context of the Kyoto Protocol, GHG emissions are not usually measured directly but derived from measurements or statistical surveys, we extend Moss and Schneider's (2000) uncertainty category "observations" to include the (not rigorously specified) category "accounting." This allows us to also consider statistically surveyed data including data (e.g., emissions data) derived with the help of statistically surveyed data (e.g., activity data) in combination with data reported in the literature (e.g., emissions factors).

The terms validation and verification, in particular, are frequently confused and misused. For instance, the IPCC Good Practice Guidelines define verification with the emphasis on GHG emissions inventories (Penman et al., 2000, p. A3.20):

Inventory definition: Verification refers to the collection of activities and procedures that can be followed during the planning and development, or after completion of an inventory that can help to establish its reliability for the intended applications of that inventory. Typically, methods external to the inventory are used to check the truth of the inventory, including comparisons with estimates made by other bodies or with emission and uptake measurements

---

[2] In the context of the Kyoto Protocol the term certification is also used, particularly by policy makers. It is specified as in Merriam-Webster (1997):
   Certification (from *certus* = certain) → certify: to attest authoritatively: to attest as meeting a standard.

determined from atmospheric concentrations or concentration gradients of these gases.

However, this definition requires discussion, as it is not sufficiently rigorously in line with either science theory or the intended purpose of the Kyoto Protocol, which may be colloquially expressed as, "It's what the atmosphere sees that matters."

According to this definition, verification is a scientific process that aims to establish the reliability of a (bottom–up) inventory. However, similar to "validity," which is a system-internal quality criterion, "reliability" is a measurement-reflexive quality criterion that should not be confused with "verification." Verification is more, as it goes beyond validation or reliability, for example, with the help of an additional experiment that allows the observation to be independently counter-checked. Moreover, in terms of checking the truth of an inventory, this definition allows "comparisons with (bottom–up emission) estimates made by other bodies"[3] to be put on the same level as "emission and uptake measurements determined from atmospheric concentrations or concentration gradients of these gases," which is unacceptable from a science–theoretical point of view, as validation and verification are confused.

2.2 Accounting Versus Diagnostic and Prognostic Modeling

Figure 2 shows the difference in terms of uncertainties between accounting and diagnostic and prognostic modeling. The accounting typically happens with a time step of ≤ 1 year and may be matched by an emission-generating model during its diagnostic mode. In its prognostic mode, a model can, at best, only reflect a multiyear period that excludes singular stochastic events (although the model may operate with a time step of ≤ 1 year).[4] The uncertainty associated with accounting U_{Account} reflects our real diagnostic capabilities. It is this uncertainty that underlies both our prior and current accounting and that, under the Kyoto Protocol, we will have to cope with in reality at some time in the future (e.g., commitment year/period). This U_{Account} may decrease

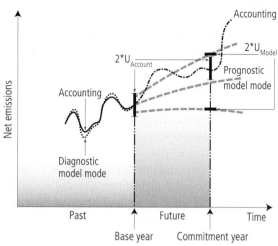

Fig. 2 Illustration of accounting versus diagnostic and prognostic modeling. *U*: uncertainty. Source: Jonas et al., (2004a, Fig. 4)

with increasing knowledge. (For simplification, we let U_{Account} stay constant in absolute terms over time in Fig. 2.) By way of contrast U_{Model}, the uncertainty of the model, always increases because of the model's decreasing prognostic capabilities with time.[5]

2.3 Uncertainty Concept

Figure 3 presents the uncertainty concept that we apply to overcome a mismatch of measured (or accounted) mean values, including their uncertainties under validation or verification. The concept acknowledges that there is both available knowledge and lack of knowledge when net carbon emissions are being accounted for. Available knowledge can be hard or soft, while lack of knowledge can be interpreted as the difference between an accepted value and the (unknown) true value that is due to unknown biases. Random errors

---

[3] In this context, the terms "third-party verification" or "independent verification" are also used.

[4] To overcome this shortcoming, stochastic events are often exogenously generated in a random fashion and introduced into prognostic models in retrospect, in the hope that their relevance will increase with respect to shorter time scales.

---

[5] The interrelation between U_{Model} and U_{Account} during the diagnostic mode of the emission-generating model can be made clear with the help of the notion of an ideal model. An ideal model perfectly reflects "reality" (inventory view) during the model's diagnostic mode, that is, U_{Model} is identical to U_{Account}. However, in practice, models are generally not able to reproduce U_{Account} for a number of reasons. An important reason is that, traditionally, model builders focused mainly on grasping mean values. To reflect more a complex reality, the models resolved more-detailed mean values. However, the consideration of uncertainties requires the opposite, that is, that models be simplified, ideally to a level that permits uncertainties to be treated as statistically independent (or as statistically independent as possible). Typically, the realization of a (sufficiently) ideal model is a task in itself.

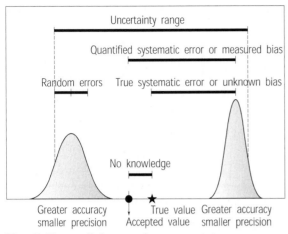

Fig. 3 The applied uncertainty concept to overcome a mismatch of measured (or accounted) mean values, including their uncertainties under validation or verification. Sources: Nilsson et al. (2000, Fig. 12), Jonas et al. (2004a, Fig. 7)

Table 1 Relative uncertainty classes applied in the full carbon account of Austria

Class	Relative uncertainty (%)
1	0–5
2	5–10
3	10–20
4	20–40
5	>40

Source: Jonas and Nilsson (2001, Section 4.1.3).

68% confidence level that the atmospheric inversion community typically applies.

2.4 Uncertainty Classes

The derivation of aggregated uncertainties, as in emission inventories, is typically not unambiguous and is even prone to errors. This is why we commonly apply relative uncertainty classes as a good practice measure (see Table 1), as they constitute a robust means of getting an effective grip on (even large) uncertainties. In light of the numerous data limitations and inconsistencies that countries face, the reporting of exact relative uncertainties is not justified.

Our work on the FCA of Austria (Jonas & Nilsson, 2001) shows that experts who share the same data sets typically estimate uncertainty ranges that overlap each other. However, this may no longer be true if the experts use different initial data, process them differently, or apply different systems views (e.g., an intramodular systems view as under partial carbon accounting (PCA) as opposed to an intermodular systems view as under FCA).[7] As a consequence of this robust finding we argue that, contrary to Gillenwater et al., (2007, Section 5), uncertainty estimates of national emission inventories can indeed be used for policy purposes. However, certain rules, particularly those dealing with large uncertainties must be obeyed (see Jonas & Nilsson, 2001, Section 4.3, for details).

Finally, we note that our definition of the relative uncertainty classes as specified in Table 1 is arbitrary and that it attempts to satisfy simple practical consid-

and systematic errors (the latter are also called determinate errors or simply biases, but we prefer quantified systematic errors or measured biases) are typically used to evaluate both hard and soft knowledge in terms of uncertainty. In contrast, lack of knowledge can only be addressed in a way that is necessary but not necessarily sufficient. This is done by defining an uncertainty range that encompasses each of the two measured biases plus each of the two standard deviations representing the random errors of the two depicted measurement sets (for comparison, see also Gillenwater et al., 2007, Section 2.2; and Winiwarter, 2007, Section 2). We note that we have not yet specified at which level of confidence we want to report uncertainty. In contrast to the IPCC (1997a, p.A1.4), which suggests the use of a 95% confidence interval, we favor the 68% confidence level (1 * standard deviation) because, as long as we have to cope with uncertainty ranges as a result of inconsistent or missing knowledge in realizing full carbon accounts, striving for a higher, purely mathematical confidence level cannot be justified physically.[6] For our discussion on bottom–up versus top–down accounting in Section 3 below, we also may want to keep in mind that it is the

---

[6]We thus distinguish between an uncertainty evaluation of Type A and Type B. Type A is the evaluation of uncertainty by the statistical analysis of a series of observations. By way of contrast, Type B is the evaluation of uncertainty by means other than the statistical analysis of series of observations (see Jonas and Nilsson, 2001, Section 4.1.2 for details).

[7]PCA as under the Revised 1996 IPCC Guidelines for National Greenhouse Gas Inventories (IPCC, 1997a, b, c) or the Kyoto Protocol do not form logical and consistent subsets of FCA (which is regarded as the scientifically appropriate approach) (Steffen et al., 1998, p.1394). However, a clear guideline on how to get from PCA to FCA, or vice versa, does not exist.

erations as to how many different intervals one wishes to resolve. The classes reflect our physical and systems analytical thinking behind Austria's full carbon account. For instance, assume that a carbon flux had been specified with a relative uncertainty of 13.7%. We then interpret this value as falling within the respective relative uncertainty class: here 10–20% (class 3).[8] In Section 5.3 below we illustrate how the concept of uncertainty classes is applied in the preparatory detection of emissions signals and the comparison of these signals across (Annex I) countries.

3 Bottom–Up Versus Top–down Accounting: Verification of Emissions

Our starting point is the verification of emissions. In this section we look at carbon emissions, the verification of which is particularly difficult (Nilsson et al., 2001; Bergamaschi, Behrend, & Jol, 2004, pp. 3–5; Nilsson et al., 2007). It requires, following science–theoretical standards, the adoption of an approach that takes an atmospheric view ("what matters is what the atmosphere sees") and is complete – leaving no unverified residues (see Fig. 4). In the context of the Kyoto Protocol, this leads us to the concept of bottom–up/top–down (consistent or dual-constrained) FCA on the country scale,[9] that is, the measurement of all fluxes, including those into and out of the atmosphere (as observed on earth), as well as an atmospheric storage measurement (as observed in the atmosphere), which – to reflect the needs of the Protocol – permits a country's "Kyoto biosphere" to be distinguished from its "non-Kyoto biosphere."[10] This type of FCA would

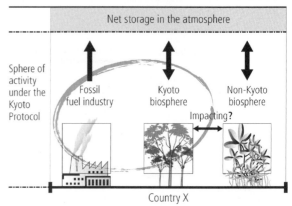

Fig. 4 PCA, as envisaged under the Kyoto Protocol (KP), must be understood as a logical subset of consistent FCA. Consistent FCA on the spatial scales of countries requires the measurement of all fluxes, including those into and out of the atmosphere, and an atmospheric storage measurement, which – to reflect the needs of the Kyoto Protocol – permits a country's "Kyoto biosphere" to be distinguished from its "non-Kyoto biosphere." The anthropogenic sector (simply referred to as fossil fuel [FF] industry) also includes ground-based fluxes between countries (e.g., trade) and carbon stocks other than biospheric stocks. Source: Jonas et al., (2004a, Fig. 5)

permit verification that is ideal because it would work both ways (bottom–up/top–down). It is, however, unattainable, as there is no atmospheric measurement available (nor likely to be in the immediate future) that can meet this discrimination requirement – not to mention the spatial (country-scale) resolution requirements of the measurement (Jonas et al., 2004a, Section 2.2; Mangino, Finn, & Scheehle, 2005: Sections 1 and 2). As a consequence, PCA – thus, partial greenhouse gas accounting (PGA) – as envisaged under the Kyoto Protocol cannot be verified.

4 Bottom–Up/Top–Down Verification of Emissions and Temporal Detection of Emissions Signals

Contrary to the bottom–up/top–down verification of emissions, however, the Kyoto Protocol requires that net emission changes (emission signals) of specified GHG sources and sinks, including those of the "Kyoto biosphere" but excluding those of the "non-Kyoto biosphere," be determined on the spatial scale of countries by the time of commitment, relative to a specified base year.[11] The relevant question then is

---

[8]The increasing width of our relative uncertainty classes and our classification of relative uncertainties as unreliable beyond class 3 is in agreement with the IPCC (1997a, p. A1.5), which advises against the application of the law of uncertainty propagation if the relative uncertainties that are combined under this law are greater than 60% (95% confidence level).

[9]The country scale is the principal reporting unit requested for reporting GHG emissions and removals under the Kyoto Protocol (FCCC, 1998, Articles 1 and 7).

[10]Articles 3.3 and 3.4 of the Protocol stipulate that human activities related to land-use change and forestry (LUCF) since 1990 can also be used to meet 2008–2012 commitments (FCCC, 1998). The part of the terrestrial biosphere that is affected by these Kyoto compliant LUCF activities is hereafter referred to as "Kyoto biosphere" and its complement as "non-Kyoto biosphere".

[11]In the figures of our paper, we denote (if not *expressis verbis*) net emissions by x and their changes by Δx, respectively.

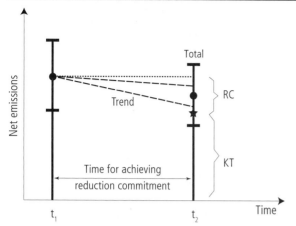

Fig. 5 The IPCC definition of uncertainty with respect to two predefined points in time (*with the respective emissions denoted by •*) based on two different types of uncertainty: total and trend uncertainty. *KT*: Kyoto emission target (*denoted by the star*); *RC*: emission reduction commitment. Source: Jonas et al., (2004a, Fig. 6)

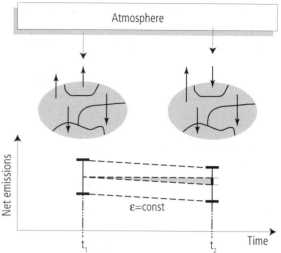

Fig. 6 Dual-constrained verification and signal detection. Source: Jonas et al., (2004a, Box 1, modified). Assume that we were able to repeatedly carry out dual-constrained FCA for a given terrestrial region at times t_1 and t_2 (appropriately averaged in space and time). Assume further that our bottom–up full carbon account would be more highly resolved than our top–down full carbon account. Nevertheless, both the bottom–up and the top–down full carbon account would exhibit "reasonable" agreement, meaning that their mean atmospheric net fluxes would be sufficiently close and could be characterized by a combined uncertainty, which would be "acceptable." However, although we would work bottom–up/top–down (i.e., apply dual-constrained FCA), we could still encounter potential difficulties, as the graph at the bottom of the figure shows. Here, the change in the net emissions at t_2 disappears within the constant-width uncertainty band. What must be kept in mind is that our bottom–up/top–down FCA technique refers to net atmospheric emissions and their uncertainties, but we need to go beyond the verification of emissions when explicitly considering time and assessing when the emission signal is outstripping uncertainty. To handle such situations, we have to additionally utilize signal detection techniques

whether these emission signals outstrip uncertainty and can be "verified" (correctly: detected).

The IPCC (to which the Kyoto Protocol appeals)[12] defines uncertainty with respect to two predefined points in time: the base year and the commitment year/period (Penman et al., 2000, Chapter 6; 2003, Chapter 5; Watson et al., 2000, Section 2.3.7). Figure 5 reflects this concept based on two different types of uncertainty, total and trend uncertainty.[13] Notwithstanding, we argue here that – if we ever want to place signal detection meaningfully into a bottom–up/top–down verification context – it is the total uncertainty in the commitment year/period that matters, as long as we are still searching for the accurate mean emission values (see Fig. 6).[14] Hence, merging bottom–up/top–

down verification of emissions and temporal detection of emission signals is the scientific challenge. It is important to realize that this challenge can be addressed successfully only if signal detection acknowledges total uncertainty. Trend uncertainty is inappropriate because it provides only second-order information (related to the change of a difference, where the difference is given by the net flux itself and the change is given by the change in the net flux over time); that is, trend uncertainty can be used to investigate how certain or uncertain an emission trend is, but it provides no information as to whether or not a realized change in net emissions is detectable.

---

[12] See FCCC (1998, Article 5; 2002, pp. 3–13; 2004, pp. 31–32).
[13] In the context of the Kyoto Protocol, the total (or level) uncertainty reflects our real diagnostic (accounting) capabilities, that is, the uncertainty that underlies our past (base year) accounting as well as our current accounting and that we will have to cope with in reality at some time in the future (commitment year/period). The trend uncertainty reflects the uncertainty of the difference in net emissions between two years (base year and/or commitment year/period).
[14] In the commitment year/period t_2 we ask, in accordance with the concept of bottom–up/top–down verification, for the total uncertainty at that point in time, not whether or not the total uncertainty at t_2 can be decreased, for example, on the basis of correlative techniques (i.e., our emission and uncertainty knowledge at t_1, the base year).

However, as discussed in Section 5.1 below, the knowledge of total uncertainty at only two points in time without a consideration of the dynamics of the emission signal can lead to interpretational difficulties as to whether or not the emission signal is detectable. (We circumvent these difficulties in Section 5.3.)

5 Temporal Detection of Emission Signals

This section focuses on the temporal detection of emission signals, which we assume to be embedded, as discussed above, in a bottom–up/top–down verification context. In Section 5.1 we explain in greater detail what we understand under a detectable emission signal vis-à-vis one that is statistically significant. Sections 5.2 and 5.3 serve to illustrate the far-reaching consequences of dealing with uncertain emission signals.

5.1 Detectability Versus Statistical Significance

Figure 7 illustrates that the notion of statistical significance is insufficient for addressing compliance under the Kyoto Protocol, as the statistical significance of an emission signal does not imply its detectability. In other words, the IPCC falls short in providing adequate support for the Protocol, as the problem of detecting emission signals – and hence, the issue of the Protocol's effectiveness (Gupta, Olsthoorn, & Rotenberg, 2003, Section 3) – still goes unresolved.[15] We address this problem with the help of the verification time (VT) concept; this perceives signal detection in the same way as climate change researchers traditionally have, that is, as a "signal-in-noise" problem (Houghton et al., 2001, Chapter 12).[16] This concept makes use of the dynamics of an emission signal and compares it with the uncertainty that underlies the emissions, not the emission signal (i.e., making the step from a to b in Fig. 8). Only a comparison of this type permits signal detection to be

---

[15]Gupta et al., (2003) argue differently but come to the same conclusion.

[16]The term "verification time" was first used by Jonas et al., (1999) and has been used by other authors since then. Actually, a more correct term is "detection time," as signal detection does not imply verification. However, we continue to use the original term, as we do not consider it inappropriate given that signal detection must, in the long term, go hand in hand with bottom–up/top–down verification.

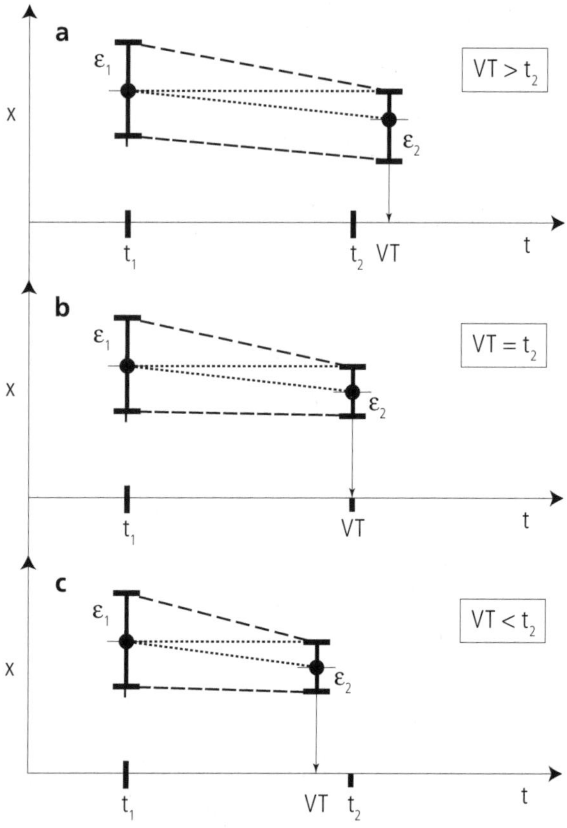

Fig. 7 Illustration of the VT concept. Assume a statistically significant (absolute) change in emissions, which outstrips uncertainty at **a** VT>t_2; **b** VT=t_2; and **c** VT<t_2. (See caption to Fig. 8 for an explanation of the symbols.) Source: Jonas et al., (2004a, Fig. 10, modified)

addressed and the question to be asked: when does an emission signal outstrip uncertainty? Considering emissions or emission changes individually within their respective uncertainty bands (i.e., staying within Fig. 8a or b, respectively) does not permit this to be done.

5.2 No Credibility Without Uncertainty

Uncertainty in the accounting matters from both a systems-analytical point of view (see Fig. 9) and an economic point of view (see Fig. 10). In Fig. 9 we study the superposition of GHG systems exhibiting different dynamics but identical effective emission signals. The figure illustrates and compares the linear and nonlinear behavior of two (here) national GHG systems in terms of their VTs. The two systems are a national anthropogenic system (simply referred to as

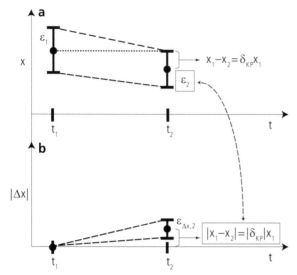

Fig. 8 **a** Emissions x_i and **b** (absolute) emission signal $|\Delta x_i|$ at t_i, together with their respective uncertainties ε_i and $\varepsilon_{\Delta x,i}$ ($i=1$, 2). δ_{KP} denotes the normalized emissions change committed to by a country under the Kyoto Protocol (see also Section 5.3). To address the question of when the emission signal outstrips uncertainty, the emission signal is compared with the uncertainty that underlies the emissions, not the emission signal (see dashed arrow between lower and upper figure). Source: Jonas et al., (2004a, Fig. A1, modified)

fossil fuel or FF system) and a national FF-plus-LUCF system. This comparison shows (see also caption to Fig. 9) that the consideration of uncertainty indeed makes a big difference in terms of the detectability of emission signals and their qualitative interpretation, even if the effective emission signals of the two countries are identical.

The same is true from an economic point of view (e.g., for emissions trading). Without uncertainty, sellers of equal amounts of carbon (or their equivalents) cannot be distinguished (Fig. 10, top), that is, they cannot be specified in terms of credibility. Figure 10 (bottom) shows that awkward cases are indeed possible, for example, when a country complying with the Kyoto Protocol performs worse than a country not complying with the Protocol. (To handle such cases requires the consideration of risk, which we do in Section 5.3.) Clearly, emissions trading can be defined in such a way that it functions according to rules that ignore uncertainties altogether, including physical scientific uncertainties. However, we doubt that this strategy will be crowned with success in the long term, especially if such rules lead to a miscon-

struction of compliance in the end and the physical scientific community thus objects to them. Hence, we argue that the success of an emissions market will crucially depend on its credibility and, thus, on the reporting of physical scientific uncertainties.

5.3 Different Techniques–Different Endings

In this section we become quantitative. We focus on the preparatory detection of emission signals, which should have been applied prior to/during the negotiation of the Kyoto Protocol. Preparatory detection allows useful information to be generated in advance regarding the possible magnitude of uncertainties due to 1) the level of confidence of the emission signal; 2) the signal one wishes to detect; and 3) the risk one is willing to tolerate in not meeting an agreed emission limitation or reduction commitment. Preparatory signal detection aims to assess emission signals in a preparatory manner, that is, at two predefined points in time: t_1 in the past/present (typically the base year) when emissions are known and t_2 in the future (typically the commitment year/period) when emissions are supposed to meet an agreed target.[17] It is correct to say that preparatory signal detection is currently more advanced in comparison with midway signal detection and signal detection in retrospect (e.g., Jonas, Nilsson, Obersteiner, Gluck, & Ermoliev, 1999; Gusti & Jęda, 2002; Dachuk, 2003; Nahorski & Jęda, 2007), Midway signal detection is carried out at some point in time between the base year and commitment year/period and considers a signal's path realized to date vis-à-vis a possible path toward the agreed emission target. Signal detection in retrospect is carried out at the end of the commitment year/period and considers how an emission signal has evolved in reality between the base year and commitment year/period.

---

[17]Different combinations of time points are referred to in the context of the Kyoto Protocol to account for GHG emissions and removals by sink and source categories on the level of countries. Without restricting generality, we use t_1 and t_2. They may refer to any two points on the time scale $T_0=1990$ (or another base year), ..., $T_{15}=2005$, ..., $T_{18}=2008$, ..., $T_{20}=2010$, ..., $T_{22}=2012$. The year 2010 is used as commitment year if t_2 refers to the temporal average in net emissions over the commitment period 2008–2012.

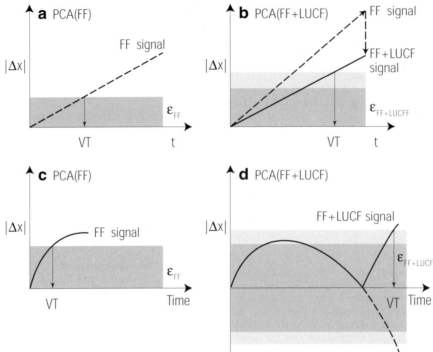

Fig. 9 Illustration of the linear (**a**, **b**) and nonlinear (**c**, **d**) behavior of VT with the help of the two partially accounted, Kyoto-eligible systems: PCA(FF) and PCA(FF+LUCF). **a, b** Here, the two systems exhibit identical effective emission signals but different uncertainties (ε_{FF} and $\varepsilon_{FF+LUCF}$, respectively, with $\varepsilon_{FF} < \varepsilon_{FF+LUCF}$) and thus different VTs. **c, d** Here, the two systems also exhibit identical effective emission signals, but now

the FF+LUCF signal exhibits a jumpy VT behavior as a result of combining a nonlinear FF signal and a LUCF signal with slow dynamics (as in **b**). (For a better overview, the LUCF signal has been omitted in **d**.) The linear and nonlinear behavior of the VT can be easily checked by slowly increasing the width of the *light-grey bar* (ε_{LUCF}), beginning from zero. Sources: Jonas and Nilsson (2001, Figs. 8, 12); see also Gusti and Jęda (2002, Fig. 17)

Our experience to date shows that there is no ideal preparatory signal detection technique; each has its pros and cons. We demonstrate this with the help of the Undershooting (Und) concept and the combined Undershooting and Verification Time (Und and VT) concept, which have been compared in detail by Jonas et al. (2004a, Sections 3.3 and 3.4). The Und concept was first described by Nahorski et al. (2003), and a more advanced version is now presented by Nahorski, Horabik and Jonas (2007), which these authors also use for their "downstream research" on the performance of carbon markets in the presence of uncertainty (see also Horabik & Nahorski, 2004).

The starting point of both the Und and the Und and VT concepts is that Annex I countries comply with their emission limitation or reduction commitments under the Kyoto Protocol.[18] They also employ the same (first-order) assumptions that

are in accordance with the preparatory signal detection concept and are fully sufficient for the purpose of this paper, viz.:

(1) Uncertainties at t_1 (base year) and t_2 (commitment year/period) are given in the form of

---

[18]For data availability reasons and because of the excellent possibility of intercountry comparisons, the Protocol's Annex I countries are used as net emitters. Their emissions/removals due to LUCF are excluded as the reporting of their uncertainties is only just becoming standard practice. The same conditions have been applied by Jonas et al., (2004b and 2004c) in their intercountry comparison of the EU member states under the EU burden sharing in compliance with the Kyoto Protocol. As a consequence of excluding emissions/removals due to land use change and forestry, our exercise here is restricted to the preparatory detection of uncertain flux signals (which we call emission signals), that is, the preparatory detection of stock-change signals is excluded. In Jonas et al., (2004a, Appendices A and C) the authors build a bridge to "stock changes" and explain how the latter can be considered.

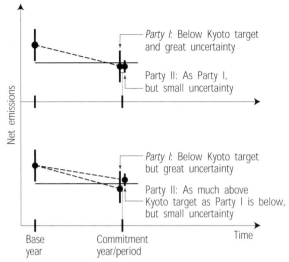

Fig. 10 Emissions trading: which country (or, more generally, "Party" in the terminology of the Kyoto Protocol) is more credible? This graphical representation illustrating the importance of uncertainty in the context of the Kyoto Protocol here addresses the crucial question of credibility while presupposing detectable net emission changes. The uncertainty intervals of both Party I and Party II encompass the same Kyoto target, but which Party is more credible in terms of emissions trading? *Top*: Both parties undershoot the Kyoto target, but Party I exhibits a greater uncertainty interval than Party II. *Bottom*: Party I exhibits a greater uncertainty interval, the mean of which undershoots the Kyoto target, while Party II exhibits a smaller uncertainty interval, the mean of which, however, does not comply with the Kyoto target

intervals that take into account that a difference might exist between the true but unknown net emissions $(x_{t,i})$ and their best estimates (x_i) ($i = 1, 2$). These differences are captured with the help of ε_i ($i = 1, 2$):

$$|x_{t,1} - x_1| \leq \varepsilon_1, \tag{1}$$

$$|x_{t,2} - x_2| \leq \varepsilon_2. \tag{2}$$

(2)　The relative uncertainty (ρ) of a country's net emissions is symmetrical and does not change over time (i.e., $\rho = $ const).

The question posed in connection with the Und concept is (see Fig. 11): by how much must countries undershoot their Kyoto targets to decrease the risk (α) that their true emissions in the commitment year/period do not undershoot (i.e., overshoot)

their true emission limitation or reduction commitments? The answer is given by:

$$x_{t,2} \geq (1 - \delta_{KP})x_{t,1} \Leftrightarrow$$

$$\frac{x_2}{x_1} \leq (1 - \delta_{KP})\frac{1 - (1 - 2\alpha)\rho}{1 + (1 - 2\alpha)\rho} \tag{3a, b}$$

$$\approx 1 - \{\delta_{KP} + 2(1 - 2\alpha)(1 - \delta_{KP})\rho\},$$

where δ_{KP} is the normalized emissions change committed by a country under the Protocol; the undershooting U is specified by:

$$U = 2(1 - \delta_{KP})\frac{(1 - 2\alpha)\rho}{1 + (1 - 2\alpha)\rho}$$

$$\approx 2(1 - 2\alpha)(1 - \delta_{KP})\rho; \tag{4a, b}$$

and the country's modified (mod) emission reduction target δ_{mod} is defined by:[19]

$$\delta_{mod} = \delta_{KP} + U. \tag{5}$$

The question posed in connection with the Und and VT concept is similar but additionally considers the detectability of an emission signal (see Fig. 12): by how much must countries undershoot their Kyoto-compatible, but detectable, targets to decrease the risk (α) that their true emissions in the commitment year/period do not undershoot (i.e., overshoot) their true emission limitation or reduction commitments? Here, the answer for the case where a country's critical (crit) or detectable emission reduction target δ_{crit} is greater than its Kyoto reduction target δ_{KP} (the case $\delta_{crit} \leq \delta_{KP}$ is covered by the Und concept above) is given by:

$$x_{t,2} \geq (1 - \delta_{crit})x_{t,1} \Leftrightarrow$$

$$\frac{x_2}{x_1} \leq (1 - \delta_{crit})\frac{1}{1 + (1 - 2\alpha)\rho}$$

$$\approx 1 - \{\delta_{KP} + U_{Gap} + (1 - 2\alpha)(1 - \delta_{crit})\rho\},$$

$$\tag{6a, b}$$

---

[19]Here, we use the Und concept in its most simple form, which does not consider any correlation between the uncertainty in the base year (ε_1) and the uncertainty in the commitment year/period (ε_2). This is a consequence of making use of the triangle inequality, which does not permit correlations to be considered. In contrast, Nahorski et al., (2007, Section 8) make use of the UND concept by applying a stochastic approach, which allows correlation to be taken into account.

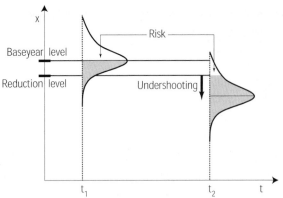

Fig. 11 Preparatory signal detection: Undershooting (Und) concept here illustrated for the case of emission reduction with the help of continuous probability distribution functions. The question posed is: how much must countries undershoot their Kyoto targets to decrease the risk of their true emissions in the commitment year/period not undershooting (i.e., overshooting) their true emission limitation or reduction commitments?

where δ_{crit}, U and U_{Gap} are specified by:

$$\delta_{\mathrm{crit}} = \frac{\rho}{1+\rho}; \tag{7}$$

$$U = U_{\mathrm{Gap}} + (1 - \delta_{\mathrm{crit}}) \frac{(1-2\alpha)\rho}{1+(1-2\alpha)\rho}$$

$$\approx U_{\mathrm{Gap}} + (1-2\alpha)(1 - \delta_{\mathrm{crit}})\rho; \tag{8a, b}$$

and

$$U_{\mathrm{Gap}} = \delta_{\mathrm{crit}} - \delta_{\mathrm{KP}}; \tag{9}$$

while the country's modified emission reduction target δ_{mod} is still given by Eq. 5.[20]

Table 2 refers to the Und concept and Table 3 to the Und and VT concept. They list the modified emission reduction targets δ_{mod} for Annex I countries committed to emission reduction, for which the "$x_{t,2}$-greater-than-$(1-\delta_{\mathrm{KP}})x_{t,1}$" or "$x_{t,2}$-greater-than-$(1-\delta_{\mathrm{crit}})x_{t,1}$" risk ($\alpha$) is specified to take on the values 0, 0.1, 0.3, and 0.5. The tables should be read as follows (compare, for example, Table 2): If a country of group 1 complies with its emission reduction commitment, that is, $x_2(=1-\delta_{\mathrm{KP}})x_1$, the risk that its true but unknown emissions $x_{t,2}$ are actually equal to or greater than its true but unknown target $(1-\delta_{\mathrm{KP}})x_{t,1}$

---

[20]The Und and VT concept only considers the uncertainty in the commitment year/period (ε_2).

is 50%. Undershooting decreases this risk. For instance, an Annex I country has committed itself to reducing its net emissions by 8%. Reporting with a relative uncertainty of $\rho=7.5\%$ (median of uncertainty class 2), the country has to reduce its emissions by 20.8% to decrease the risk from 50 to 0%.

Table 2 shows that the Und concept is difficult to justify politically in the context of the Kyoto Protocol. Under the Protocol, nonuniform emission reduction commitments (see δ_{KP} values in the third column) were determined "off the cuff," meaning that they were derived via horse trading and not as a result of rigorous scientific considerations. The outcome is discouraging. Varying δ_{KP} while keeping the relative uncertainty (ρ) and the risk (α) constant shows that Annex I countries complying with a smaller δ_{KP} are better-off than countries that must comply with a greater δ_{KP}. (See, for example, the bolded δ_{mod} values in the column for $\rho=7.5\%$, which refer to the same risk $\alpha=0.3$ and decrease with decreasing δ_{KP}.) Such a situation is not in line with the spirit of the Kyoto Protocol.

Table 3, on the other hand, reveals crucial difficulties in terms of realizing the Und and VT concept. This concept requires the Protocol's emission reduction targets for nondetectability to be corrected through the introduction of an initial or obligatory undershooting (U_{Gap}) so that the countries' emission signals become detectable (i.e., meet the maximal allowable VT) before the countries are permitted to make economic use of their excess emission reductions. (See, for example, group 1 countries; that is, the line for $\delta_{\mathrm{KP}}=8\%$: the δ_{mod} value for $\rho=15\%$ (median of uncertainty class 3) and $\alpha=0.5$ is $\delta_{\mathrm{mod}} = \delta_{\mathrm{KP}} +$

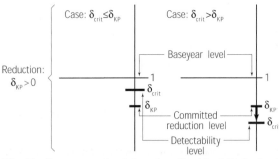

Fig. 12 Preparatory signal detection: Combined Undershooting and Verification Time (Und&VT) concept here for the case of emission reduction. Here the relevant question, though similar to the one posed under the Und concept, additionally considers the detectability of emission signals

mates. *Water, Air, & Soil Pollution: Focus* (in press) doi:10.1007/s11267-006-9114-6.

Nahorski, Z., Jęda, W., & Jonas, M. (2003). Coping with uncertainty in verification of the Kyoto obligations. In J. Studzinski, L. Drelichowski, & O. Hryniewicz (Eds.), *Zastosowanai Informatyki i Analizy Systemowej w Zarzadzaniu* (pp. 305–317). Warsaw, Poland: Systems Research Institute of the Polish Academy of Sciences.

Nahorski, Z., Horabik, J., & Jonas, M. (2007). Compliance and emissions trading under the Kyoto protocol: Rules for uncertain inventories. *Water, Air, & Soil Pollution: Focus* (in press) doi:10.1007/s11267-006-9112-8.

Nilsson, S., Jonas, M., & Obersteiner, M. (2002). COP 6: A healing shock? *Climatic Change, 52*(1–2), 25–28.

Nilsson, S., Jonas, M., Obersteiner, M., & Victor, D. G. (2001). Verification: The gorilla in the struggle to slow global warming. *Forestry Chronicle, 77*(3), 475–478.

Nilsson, S., Shvidenko, A., Jonas, M., McCallum, I., Thompson, A., & Balzter, H. (2007). Uncertainties of the regional terrestrial biota full carbon account: A systems analysis. *Water, Air, & Soil Pollution: Focus* (in press) doi:10.1007/s11267-006-9119-1.

Nilsson, S., Shvidenko, A., Stolbovoi, V., Gluck, M., Jonas, M., & Obersteiner, M. (2000). Full carbon account for Russia. *Interim Report IR-00-021*, International Institute for Applied Systems Analysis, Laxenburg, Austria. Available at: http://www.iiasa.ac.at /Publications/Documents/IR-00-021.pdf.

Penman, J., Kruger, D., Galbally, I., Hiraishi, T., Nyenzi, B., Emmanuel, S., et al. (Eds.) (2000). *Good practice guidance and uncertainty management in national greenhouse gas inventories*. Hayama, Kanagawa, Japan: Institute for Global Environmental Strategies. Available at: http://www.ipcc-nggip.iges.or.jp/public/gp/english/.

Penman, J., Gytarsky M., Hiraishi, T., Krug, T., Pipatti, D., Buendia, R. L., et al. (Eds.) (2003). *Good practice guidance for land use, land-use change and forestry*. Hayama, Kanagawa, Japan: Institute for Global Environmental Strategies. Available at: http://www.ipcc-nggip.iges.or.jp/public/gpglulucf /gpglulucf.htm.

Shvidenko, A., & Nilsson, S. (2003). A synthesis of the impact of Russian forests on the global carbon budget for 1961–1998. *Tellus, 55B*, 391–415.

Schulze, E.-D., Valentini, R., & Sanz, M.-J. (2002). The long way from Kyoto to Marrakesh: Implications of the Kyoto protocol negotiations for global ecology. *Global Change Biology, 8*, 505–518.

Steffen, W., Noble, I., Canadell, J., Apps, M., Schulze, E.-D., Jarvis, P. G., et al. (1998). The terrestrial carbon cycle: Implications for the Kyoto protocol. *Science, 280*, 1393–1394.

Valentini, R., Dolman, A. J., Ciais, P., Schulze, E.-D., Freibauer, A., Schimel, D., et al. (2000). *Accounting for carbon sinks in the biosphere, European perspective*. Jena, Germany: CarboEurope European Office, Max-Planck-Institute for Biogeochemistry. (Listed under: http://www.bgc-jena.mpg.de/public/carboeur/).

Vreuls, H. H. J. (2004). Uncertainty analysis of Dutch greenhouse gas emission data, a first qualitative and quantitative (TIER 2) analysis. In *Proceedings, international workshop on uncertainty in greenhouse gas inventories: Verification, compliance and trading, held 24–25 September, Warsaw, Poland*, 34–44. Available at: http://www.ibspan.waw.pl/GHGUncert2004/ schedule.htm.

Watson, R. T., Noble, I. R., Bolin, B., Ravindranath, N. H., Verardo, D. J., & Dokken, D. J. (Eds.) (2000). *Land use, land-use change, and forestry*. Cambridge, UK: Cambridge University Press. Available at: http://www.grida.no/climate/ipcc/land_use/.

Winiwarter, W. (2007). National greenhouse gas inventories: Understanding uncertainties versus potential for improving reliability. *Water, Air, & Soil Pollution: Focus* (in press) doi:10.1007/s11267-006-9117-3.

Water Air Soil Pollut: Focus (2007) 7:513–527
DOI 10.1007/s11267-006-9114-6

Processing National CO$_2$ Inventory Emissions Data and their Total Uncertainty Estimates

Zbigniew Nahorski · Waldemar Jęda

Received: 9 May 2005 / Accepted: 20 June 2006 / Published online: 3 February 2007
© Springer Science + Business Media B.V. 2007

Abstract The uncertainty of reported greenhouse gases emission inventories obtained by the aggregation of partial emissions from all sources and estimated to date for several countries is very high in comparison with the countries' emissions limitation and reduction commitments under the Kyoto Protocol. Independent calculation of the estimates could confirm or question the undertainty estimates values obtained thus far. One of the aims of this paper is to propose statistical signal processing methods to enable calculation of the inventory variances. The annual reported emissions are used and temporal smoothness of the emissions curve is assumed. The methods considered are: a spline-function-smoothing procedure; a time-varying parameter model; and the geometric Brownian motion model. These are validated on historical observations of the CO$_2$ emissions from fossil fuel combustion. The estimates of variances obtained are in a similar range to those obtained from national inventories using TIER1 or TIER2. Additionally, some regularities in the observed curves were noticed.

Keywords modelling CO$_2$ emissions · nonparametric methods · parametric methods · geometric Brownian motion · estimation of variance

1 Introduction

Under the Kyoto Protocol (FCCC, 1998, 2001), Parties have an obligation to decrease their greenhouse gas emissions by 5.2% below 1990 levels by 2008–2012. The greenhouse gas emission inventories of each country are monitored by the Secretariat of the United Nations Framework Convention on Climate Change. However, in the majority of national accounts, the uncertainty ranges exceed, sometimes very considerably, the emission reductions agreed upon in Annex I to the Protocol (see Winiwarter, 2007).

The signatories to Annex I must monitor their emissions starting from the base year which, for most countries, is 1990. A dozen emission inventories for each country are thus already available (see e.g. Brandes, Olivier, & Oorschot 2004; Rypdal & Winiwarter 2001). These could perhaps be useful for improving the estimates of individual emissions in the commitment period 2008–2012, using statistical inference. They could also be used to estimate the parameters of the statistical distribution of the inventory errors, thereby providing independent assessment of the

Z. Nahorski (✉) · W. Jęda
Systems Research Institute,
Polish Academy of Sciences, Newelska 6,
PL-01-447 Warsaw, Poland
e-mail: Zbigniew.Nahorski@ibspan.waw.pl

🖄 Springer

range of errors estimated to date by the prop- agation of the initial errors in the calculations. Both these tasks are addressed in this paper. The methods used are validated for data on carbon dioxide (CO_2) emissions from fossil fuel combus- tion estimated for the years 1851–1998 (Marland, Boden, & Andres, 2006).

There are a handful of methods that can be used for data smoothing. We use three meth- ods for estimating emissions and their variances that we consider to be particularly well suited to the above problem: smoothing splines (Wahba, 1990; Gu, 2002); a parametric model with a time-variable coefficient; and the Brownian mo- tion model. Other methods could also be tried with a view to solving the problem, for exam- ple another smoothing method. Methods based on wavelets might be promising (Debnath, 2002; Walter, 1994). Moreover, fitting by a set of poly- nomials that do not satisfy the smoothing condi- tions required for the spline methods may give promising results. Popular methods in the auto- matic control literature use parametric models with calculation of the state errors following an earlier phase of parameter estimation. In some of them, for instance, the extended Kalman filter, the parameters and the states are estimated simul- taneously. To use such methods, the parametric model is needed. Otherwise, most require quite long data samples to converge.

The projection of emissions into the fu- ture is of interest in many investigations (see Kroeze, Vlasblom, Gupta, Boudri, & Blok, 2004; Manne & Richels, 2004; Riahi, Rubin, Taylor, Schrattenholzer, & Houndshell, 2004), as it is con- nected, for example, with anticipation of the na- tional greenhouse gas balances in the commitment period or, on the global scale, with anticipation of supply and demand to predict the price of tradable emissions permits. Although the results presented in this paper could also be used for projection purposes, the paper concentrates more on histori- cal data and on the possibility of gathering useful information from them.

Thus, we estimate the variance of the errors in the reported CO_2 emissions, thereby obtaining independent values of the uncertainty level. Al- though the uncertainties considered here are not exactly the same as those estimated as impreci-

sions in inventories, the estimates turn out to be in a similar range for both cases. This supports, in a sense, the correctness of the methods applied to estimate the inventory uncertainties to date, at least for fossil fuel emissions.

To estimate the variance, different methods of modeling the emissions were used (see IPCC, 1996; IPCC, 2000). When they were applied to the historical data, some regularities were noticed. The emissions often follow piecewise exponential curves, particularly in periods of steady growth. Much less regular data are observed in the decline periods, emissions becoming highly irregular in times of war and during the transition from growth to decline.

Problems connected with greenhouse gas in- ventory uncertainty were signaled in many pa- pers (e.g. Gupta, Oltshoorn, & Rotenberg, 2003; Monni, Syri, Pipatti, & Savolainen, 2007; Nilsson, Jonas, Obersteiner, & Victor, 2001; Nilsson, Jonas, & Obersteiner, 2002; Nahorski, Jeda, & Jonas, 2003). Proper estimation of its level is therefore of major importance.

In Section 2 a basic notation is introduced. Section 3 presents the nonparametric method based on smoothing splines. In Section 4 the appli- cation of a parametric method is discussed and some numerical results are presented. Section 5 briefly discusses the possibility of using the Brownian motion model to describe the evolu- tion of national CO_2 emissions in time. Section 6 concludes.

2 Notation Used

By $x(t)$ as a function of time, we denote the inte- gral of the real emissions calculated on the inter- val $(t - 1, t]$, where t is expressed in years. Thus, the integral is calculated over the one-year-back period. In the sequel we call $x(t)$ the emissions. The function $x(t)$, as the integral of a positive function, is continuous and positive. In this paper we assume that $x(t)$ is a smooth enough function. The emissions balances provided by the Annex I Parties are prepared by making an inventory of the emissions from all relevant activities during a year. Because of uncertainties in assessing the ex- act quantities and coefficients, they include errors.

evolution of the data. We then explain why the model was chosen and finally present some results for fitting the model to the emissions data for some countries.

As we assumed that x_i are positive, we can define a new time series

$$g_i = \frac{x_{i+1}}{x_i} - 1 = \frac{x_{i+1} - x_i}{x_i}, \qquad i = 0, 1, \ldots, N - 1.$$

Each element g_i of a new time series can be interpreted as a relative difference of the two consecutive elements x_{i+1} and x_i.

From the latter relation we can now formulate the following difference equation

$$x_{i+1} - x_i = g_i x_i, \qquad x_0 = x(t_0). \qquad (7)$$

Because $y_i = (1 + u_i)x_i$, then Eq. 7 can be transformed to

$$y_{i+1} = (1 + g_i)\frac{1 + u_{i+1}}{1 + u_i} y_i.$$

Dividing both sides by y_0 and taking logarithms yields

$$Y_{i+1} = \ln(1 + g_i) + \ln \frac{1 + u_{i+1}}{1 + u_i} + Y_i$$

or approximately

$$Y_{i+1} - Y_i \approx g_i + u_{i+1} - u_i,$$

from which an estimator $\hat{g}_i$ can be designed as

$$\hat{g}_i = Y_{i+1} - Y_i. \qquad (8)$$

Under our assumption on values of u_i we have

$$E(\hat{g}_i) = E(Y_{i+1} - Y_i + u_i - u_{i+1})$$
$$= X_{i+1} - X_i = \ln(1 + g_i) \approx g_i.$$

Thus, the estimator is approximately unbiased. Its approximate variance is

$$\mathrm{var}(\hat{g}_i) = E(Y_{i+1} - X_{i+1} - Y_i + X_i)^2 =$$
$$= E(u_{i+1} - u_0 - u_i + u_0)^2 = E(u_{i+1} - u_i)^2$$
$$= \sigma_{i+1}^2 - 2\gamma_{i,i+1} + \sigma_i^2,$$

from which the estimate of the standard deviation $\hat{\sigma}_{\hat{g}_i}(N)$ can be obtained.

4.1 Estimation of the Parameter g_i

The expression (8) was used to estimate the function g_i for a few countries from CO_2 emissions data from fossil fuel combustion data mentioned above (Marland et al., 2006). Some chosen results are presented in Figs. 2, 3, 4, and 5. The smoothing splines were used to smooth the points obtained from Eq. 8 with the formulae (3). For each country, the reported emissions (dots) and their smoothing spline approximations (solid lines) are depicted in the left panel. The right panel shows estimates of the function g_i. The dots represent the points calculated using the formula (8). The bold solid line is obtained by smoothing these

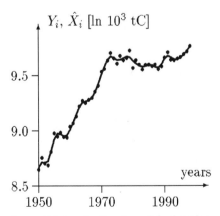
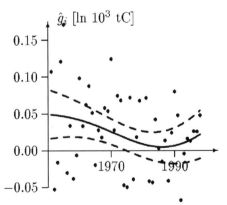

Fig. 3 Results of smoothing and estimation of the function g for Austria in the years 1950–1998. *Left panels*: dots – logarithms of reported emissions, *solid lines* – smoothed logarithms of reported emissions. *Right panels*: dots – esti-

mates of $\hat{g}_i$ from the formula (8), the *solid bold lines* – their smoothed continuous approximations, the *dashed lines* of normal thickness – the 95% confidence intervals of these approximations

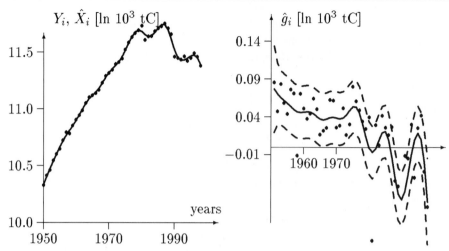

Fig. 4 Results of smoothing and estimation of the function g for Poland in the years 1950–1998. *Left panels: dots –* logarithms of reported emissions, *solid lines –* smoothed logarithms of reported emissions. *Right panels: dots –* esti-mates of $\hat{g}_i$ from the formula (8), the *solid bold lines –* their smoothed continuous approximations, the *dashed lines* of normal thickness – the 95% confidence intervals of these approximations

points. The dashed lines show the 95% confidence intervals of the estimates.

Table 1 also depicts the estimates of the standard deviation of the errors $u_{i+1} - u_i$. By comparing these with the values obtained from smoothing, it can be seen that both estimates of the standard deviations are of the same order, although not always very close to each other. Note, however, that the values from smoothing corre-spond to the standard deviations of the errors $u_i - u_0$, while those from the parametric model correspond to $u_i - u_{i-1}$, which might partly cause the differences.

4.2 Piecewise Exponential Model

Although the estimated functions $\hat{g}(t)$ in the pre-vious section vary in time, in many periods their patterns resemble the constant value lines. To bet-ter investigate this question let us start by examin-ing a few curves. Figures 6 and 7 contain emission curves y_i and logarithmic curves $Y_i = \ln(y_i/y_0)$, $t_0 = 1990$, for the emission data (Marland et al., 2006) for Australia and the United States. It can be seen that the data evolve approximately along a piecewise exponential curve and that the loga-rithmic curves are approximately linear.

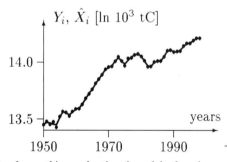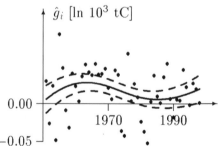

Fig. 5 Results of smoothing and estimation of the function g for USA in the years 1950-1998. *Left panels: dots –* logarithms of reported emissions, *solid lines –* smoothed logarithms of reported emissions. *Right panels: dots –* esti-mates of $\hat{g}_i$ from the formula (8), the *solid bold lines –* their smoothed continuous approximations, the *dashed lines* of normal thickness – the 95% confidence intervals of these approximations

Fig. 6 Emissions for Australia with fitted piecewise exponential curve (*left*) and their logarithms with fitted straight lines (*right*), in millions of metric tons of C

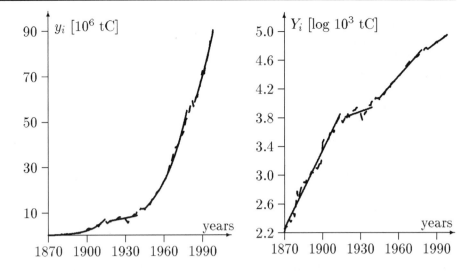

Thus, the exponential growth models describe data development adequately only in some definite intervals. These intervals are the periods of constant development conditions. One can easily distinguish in the figures the period of the eighteenth-century industrial revolution or the period of postwar prosperity of the 1950s–1970s. However, even for the United States (Fig. 7), and more visibly for European countries like Poland or Austria (Figs. 8 and 9), it is easy to see that there are periods where the assumption regarding the simple constant parameter g (and therefore the growth along the exponential curve) cannot be true. This is particularly visible in the world-war periods, the 1930s depression years, and the energy shocks of the 1970s–1980s. Smaller ripples can be distinguished and explained, as, for exam-

ple, in the case of the Polish transformation period. The fit of this simple piecewise exponential model is quite good in periods of growth or decay. In a period of steady growth it is almost perfect. In the periods of decline, the emissions are often more volatile. War and transition periods, like those of 1970s in western Europe or of the 1980s in Poland, are highly irregular and were skipped from fitting.

The results obtained are generally quite similar for both methods. The error variance estimates calculated by the regression method (piecewise exponential model) usually turn out to be greater than those calculated in Section 3 (Table 1). This seems to be connected with oversimplification in the exponential model used, as seen in Figs 10, 11, 12 and 13, where the estimates of g from the

Fig. 7 Emissions for United States with fitted piecewise exponential curve (*left*) and their logarithms with fitted straight lines (*right*), in millions of metric tons of C

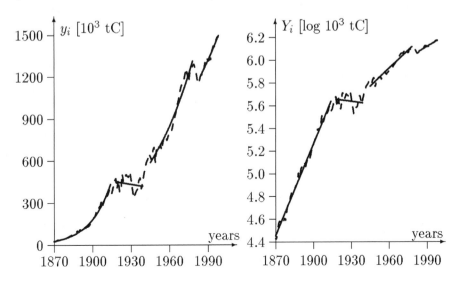

Fig. 8 Emissions for Poland with fitted piecewise exponential curve (*left*) and their logarithms with fitted straight lines (*right*), in millions of metric tons of C

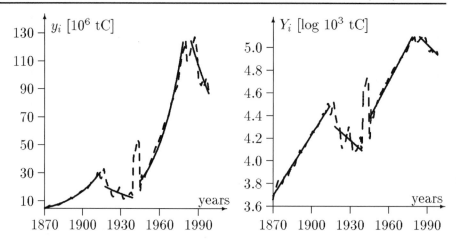

smoothing and from the piecewise exponential models are compared. However, the good fit of the piecewise exponential model, although to be expected in theory, seems to be an important observation. This means that in the past the emissions have approximately followed the exponential functions in defined longer periods. The jump from one such segment to another is for the most part connected with a big political or economic change.

5 Geometric Brownian Motion

Geometric Brownian motion is the most-often-used stochastic process in financial economics theory, and in our case may be considered as a useful alternative from the practical point of view. In several cases it was not found to be a better model

than others, even though it provides reasonable mapping of probabilities in time.

5.1 Geometric Brownian Model for the Emissions

For a stochastic process $x(t)$ that follows a geometric Brownian motion, the stochastic equation for its variation in time t is

$$dx = gxdt + \sigma xdz, \tag{9}$$

where $dz = \varepsilon dt^{1/2}$ is the Wiener increment, ε is a standard normally distributed random variable, g is the drift, and σ is the volatility of x.

In the above equation the first term on the right-hand side is the expectation (trend) term and the second term is the variation term (deviation from the trend or uncertainty).

Fig. 9 Emissions for Austria with fitted piecewise exponential curve (*left*) and their logarithms with fitted straight lines (*right*), in millions of metric tons of C

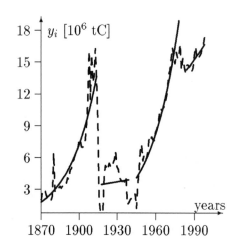
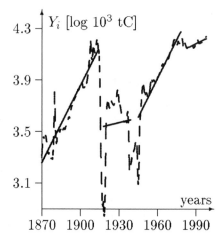

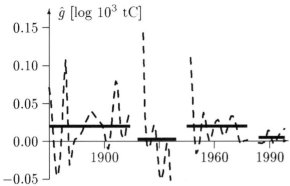

Fig. 10 Estimates of g for Austria. *Solid line* – piecewise exponential model, *dashed lines* – smoothing

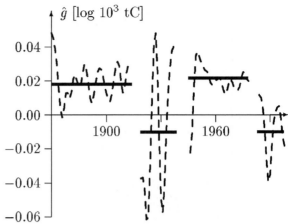

Fig. 12 Estimates of g for Poland. *Solid line* – piecewise exponential model, *dashed line* – smoothing

The geometric Brownian motion is a log-normal diffusion process with the expected value of x at the time t (starting at $t_0 = 0$)

$$E[x(t)] = x_0 e^{gt} \tag{10}$$

and the standard deviation SD

$$SD[x(t)] = x_0 e^{gt} \sqrt{e^{\sigma^2 t} - 1}. \tag{11}$$

This is illustrated in Fig. 14.

5.2 Arithmetic Brownian Model for the Logarithm of the Emissions

Because of its simplicity, it is useful to work with the logarithmic diffusion equation. Letting $X =$

$\ln x$, and using Itô's lemma we find that x follows the arithmetic (or ordinary) Brownian motion

$$dX = d\ln x = 2\left(g - \frac{1}{2}\sigma^2\right)dt + \sigma\,dz, \tag{12}$$

so

$$dX = g'dt + \sigma\,dz,$$

where $g' = 2\left(g - \frac{1}{2}\sigma^2\right)$. The variable X follows an arithmetic Brownian motion with the drift g' and volatility σ.

We should note here that although the volatility term is the same in Eq. 12 as in the geometric

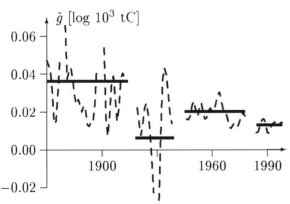

Fig. 11 Estimates of g for Australia. *Solid line* – piecewise exponential model, *dashed line* – smoothing

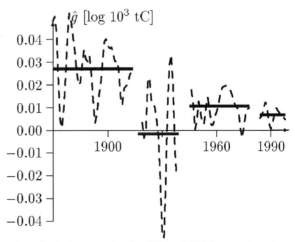

Fig. 13 Estimates of g for USA. *Solid line* – piecewise exponential model, *dashed line* – smoothing

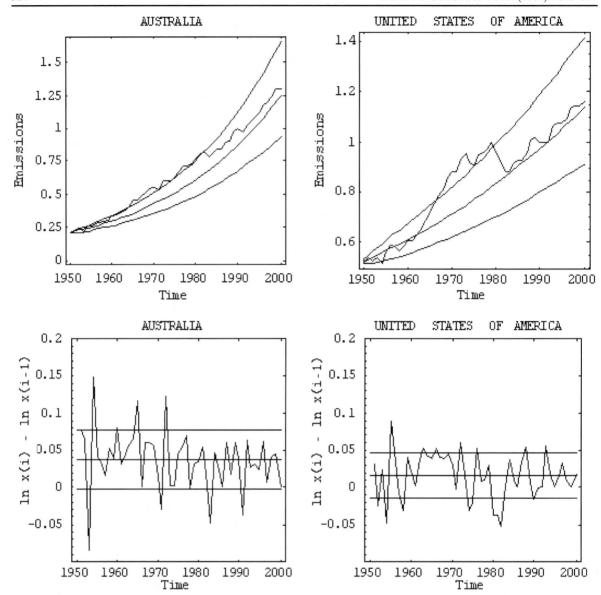

Fig. 14 Illustrations of the stochastic process considered, showing sample paths, the 66% confidence intervals, and the forecasted expected values (exponential trend line) for two countries (*upper panels*) and estimates $\hat{g}_i = Y_{i+1} - Y_i$ with their means and 66% confidence intervals (*lower panels*)

Brownian motion for x (9), the element $d(\ln x)$ is different from dx/x because of the different drift expression (the so-called Itô's effect).

The drift parameter g can be estimated as the average value of a set of differences of the logarithms $\ln y_{i+1} - \ln y_i = Y_{i+1} - Y_i \approx \hat{g}_i$. Using the same historical series we can obtain an estimation of the volatility σ by taking the standard deviation of $Y_{i+1} - Y_i$, as for the parametric model of Section 4. These can be inserted in Eqs. 10 and 11 to obtain the characterization of the process in time.

The calculations, not presented here, give bigger estimates of the standard deviations than those depicted in Table 1, which are comparable to the piecewise exponential model of Section 4. These bigger values seem to be caused mainly by the constant value of g in the model.

Table 3 Comparison of methods applied in the paper

Method	Spline smoothing	Variable parameter	Piecewise smoothing	Brownian motion
Assumptions	Uncorrelated $v_i = u_i - u_0$ u_i – small relative zero mean error	u_i – small relative zero mean error	u_i – small relative zero mean error	εdt – Wiener process
Theoretical properties	Consistent estimator of variance	Approximately unbiased parameter estimator	?	?
Data length	$N > 30$	Any, but $N > 30$ if smoothed	Any	?
Accuracy	Very good	Very good	Middle	Middle
Robustness	Middle	Middle	High, except turning points	Low
Application	Estimation of mean and variance	Estimation of mean and variance	Limited prognosis	Prognosis
Applicability	Difficult	Easy	Easy	Difficult

Further elaboration of the Brownian motion method is beyond the scope of this paper and should be covered in a separate paper.

6 Conclusions

Nonparametric and parametric methods for modeling phenomena of greenhouse gas emissions and for estimating the parameters are proposed in this paper. These differ in terms of the degree of smoothing and precision of fitting of the reported emissions. A comparison of the methods used (see also Table 3) reveals that the parametric method of Section 4 in many instances gives simpler, less volatile curves, although it is more sensitive to the smoothing interval. The smoothing method of Section 3 is more accurate and better emphasizes the ripples in the data. The parametric piecewise exponential model gives the roughest but also the simplest description, showing general trends in evolution of emissions data.

One of the main goals of the paper was to estimate the standard deviation of the errors.

Some signal processing methods are proposed, and preliminary results are presented. These are based on the published estimates of emissions from fossil fuel combustion (Marland et al., 2006) and therefore do not cover the entire greenhouse gas emissions spectrum reported within the Kyoto agreement. Moreover, the volatility of reported emissions may be related not only to observation errors but also to such factors as changing weather conditions and a country's rapidly changing economic situation. These phenomena might have contributed to an increase in the estimated variance. It is also worth remembering that the variance is estimated with some accuracy.

With these reservations, the calculations performed for fossil fuels indicate that the empirical approach gives reasonable estimates, comparable to the estimates obtained to date by the methods recommended by the Intergovernmental Panel on Climate Change (IPPC) (IPCC, 2000). Or, to be more cautious, the partial results obtained here do not falsify the uncertainty estimation procedures applied to date for the inventories. It is, however, impossible at this stage to draw any definite conclusions as to the accuracy of uncertainty

Springer

estimates of greenhouse gases where the scope of the expert knowledge used is much wider.

An interesting relation between the piecewise exponential growth of the CO_2 emissions curve and a country's economic development may well be true for other gases as well. An open question is how the removal of greenhouse gases by sinks, also included in the full calculation of the greenhouse gas balance of countries, may behave. It will be possible, in time, as longer historical records become available, to analyze the evolution of this type of data. The proposed approach can be used to better estimate the real emissions by filtering out errors, and possibly for the purpose of prognosis. The latter application could be important as an alternative to the scenarios built on the basis of technological and economic assumptions. But such an application is still quite risky until more is known about the dependence of emissions on economic, structural, or weather conditions.

Acknowledgements The authors are indebted to Matthias Jonas of IIASA, Laxenburg, Austria, for his comments on an earlier version of the paper. Partial financial support from the Polish State Scientific Research Committee within the grant 3PO4G12024 is gratefully acknowledged.

References

Brandes, L. J., Olivier, J. C. J., & van Oorschot, M. H. P. (2004). Validation, verification and uncertainty assessment for improving The Netherlands' emission inventory. In *Proceedings of the workshop 'Uncertainty in Greenhouse Gas Inventories: Verification, Compliance & Trading.'* (pp. 19–33). Warsaw, Poland: SRI PAS & IIASA. http://www.ibspan.waw.pl/papers/Brandes.pdf.

Debnath, L. (2002). *Wavelet transforms and their applications*. Switzerland: Birkhäuser.

FCCC (1998). *FCCC report of the conference of the parties on its third session, held at Kyoto from 1 to 11 December 1997*. Addendum. Document FCCC/CP/1997/7/Add.1, United Nations Framework Convention on Climate Change (FCCC). http://www.unfccc.de/ http://unfccc.int/index.html.

FCCC (2001). Implementation of the Buenos Aires plan of action: adoption of the decisions giving effect to the Bonn agreements. Draft decisions forwarded for elaboration, completion and adoption. National systems, adjustments and guidelines under Articles 5, 7 and 8 of the Kyoto Protocol. Document FCCC/CP/2001/L.18, United Nations Framework Convention on Climate Change (FCCC). http://www.unfccc.de/.

Gillenwater, M., Sussman, F., & Cohen, J. (2007). Practical applications of uncertainty analysis for national greenhouse gas inventories (this issue).

Gu, C. (2002). *Smoothing spline ANOVA models*. Berlin Heidelberg New York: Springer.

Gugele, B., Huttunen, K., & Ritter, M. (2005). *Annual European community greenhouse gas inventory 1990–2003 and inventory report 2005*. Technical report no. 4/2005. Copenhagen, Denmark: European Environment Agency. http://reports.eea.europa.eu/technical_report_2005_4/en.

Gupta, J., Oltshoorn, X., & Rotenberg, E. (2003). The role of scientific uncertainty in compliance with the Kyoto Protocol to the climate change convention. *Environmental Science and Policy, 6*, 475–486.

Hudz, H. (2003). *Verification times underlying the Kyoto Protocol: Consideration of risk.* Interim report IP-02-066. Austria: International Institute for Applied Systems Analysis (*IIASA*).

IPCC (1996). *IPCC guidelines for national greenhouse gas inventories, vol. 1–3*. London: IPCC.

IPCC (2000). Good practice guidance and uncertainty management in national greenhouse gas inventories. In J. Penman, D. Kruger, I. Galbally, T. Hiraishi, B. Nyenzi, S. Emmanuel, L. Buendia, R. Hoppaus, T. Martinsen, J. Meijer, K. Miwa, & K. Tanabe (Eds.), *Intergovernmental panel on climate change (IPCC) National gas inventories program*. Technical support unit. Hayama, Kanagawa, Japan: Institute for Global Environmental Strategies.

Jonas, M., & Nilsson, S. (2007). Prior to an economic treatment of emissions and their uncertainties under the Kyoto Protocol: scientific uncertainties that must be kept in mind (this issue).

Kroeze, C., Vlasblom, J., Gupta, J., Boudri, C., & Blok, K. (2004). The power sector in China and India: greenhouse gas emission potential and scenarios for 1990–2020. *Energy Policy, 32*, 55–76.

Manne, A., & Richels, R. (2004). US rejection of the Kyoto Protocol: the impact on compliance cost and CO_2 emissions. *Energy Policy, 32*, 447–454.

Marland, G., Boden, T. A., & Andres, R. J. (2006). Global, regional, and national fossil fuel CO_2 emissions. In *Trends: A compendium of data on global change, carbon dioxide information analysis center*. OAK Ridge National Laboratory: U.S. Department of Energy. http://cdiac.esd.ornl.gov/trends/emis/em_cont.htm.

Monni, S., Syri, S., Pipatti, R., & Savolainen, I. (2007). Extension of EU emissions trading scheme to other sectors and gases: consequences for uncertainty of total tradable amount (this issue).

Nahorski, Z., Jęda, W., & Jonas, M. (2003). Coping with uncertainty in verification of the Kyoto obligations. In J. Studziński, L. Drelichowski, & O. Hryniewicz (Eds.), *Zastosowania informatyki i analizy systemowej w zarządzaniu*, (pp. 305–317). Warsaw, Poland: IBS PAN.

Nilsson, S., Jonas, M., Obersteiner, M., & Victor, D. (2001). Verification: the gorilla in the struggle to slow global warming. *Forestry Chronicle, 77*, 475–478.

Nilsson, S., Jonas, M., & Obersteiner, M. (2002). COP 6: a healing shock. *Climatic Change, 52*, 25–28.

Riahi, K., Rubin, E. S., Taylor, M. R., Schrattenholzer, L., & Houndshell, D. (2004). Technological learning for carbon capture and sequestration technologies. *Energy Policy, 26*, 539–564.

Rypdal, K., & Winiwarter, W. (2001). Uncertainty in greenhouse gas emission inventories – evaluation, comparability and implications. *Environmental Science and Policy, 4*, 104–116.

Wahba, G. (1990). *Spline models for observational data.* Montpelier, VT, USA: Capital City Press.

Walter, G. G. (1994). *Wavelets and other orthogonal systems with applications.* CRC Press.

Winiwarter, W. (2007). *National greenhouse gas inventories: understanding uncertainties versus potential for improving reliability* (this issue).

Water Air Soil Pollut: Focus (2007) 7:529–538
DOI 10.1007/s11267-006-9111-9

Extension of EU Emissions Trading Scheme to Other Sectors and Gases: Consequences for Uncertainty of Total Tradable Amount

S. Monni · S. Syri · R. Pipatti · I. Savolainen

Received: 28 May 2006 / Accepted: 20 December 2006 / Published online: 23 January 2007
© Springer Science + Business Media B.V. 2007

Abstract Emissions trading in the European Union (EU), covering the least uncertain emission sources of greenhouse gas emission inventories (CO_2 from combustion and selected industrial processes in large installations), began in 2005. During the first commitment period of the Kyoto Protocol (2008–2012), the emissions trading between Parties to the Protocol will cover all greenhouse gases (CO_2, CH_4, N_2O, HFCs, PFCs, and SF_6) and sectors (energy, industry, agriculture, waste, and selected land-use activities) included in the Protocol. In this paper, we estimate the uncertainties in different emissions trading schemes based on uncertainties in corresponding inventories. According to the results, uncertainty in emissions from the EU15 and the EU25 included in the first phase of the EU emissions trading scheme (2005–2007) is ±3% (at 95% confidence interval relative to the mean value). If the trading were extended to CH_4 and N_2O, in addition to CO_2, but no new emissions sectors were included, the tradable amount of emissions would increase by only 2% and the uncertainty in the emissions would range from −4 to +8%. Finally, uncertainty in emissions included in emissions trading under the Kyoto Protocol was estimated to vary from −6 to +21%. Inclusion of removals from forest-related activities under the Kyoto Protocol did not notably affect uncertainty, as the volume of these removals is estimated to be small.

Keywords emissions trading · EU · greenhouse gas · Kyoto Protocol · uncertainty

1 Introduction

In the 2008–2012 commitment period, the Kyoto Protocol aims to reduce greenhouse gas emissions from industrial countries by an average of 5% from the 1990 level. Several flexibility mechanisms (emissions trading, joint implementation, and the clean development mechanism) have been implemented to lower the overall costs of achieving emission targets.

The European Union (EU) started its carbon dioxide (CO_2) emissions trading scheme (EU ETS) in 2005 both to improve cost-efficiency in emission reductions and to give member states experience in emissions trading (Official Journal of the European Union, 2003). CO_2 emissions from power plants with thermal capacity greater than 20 MW, and emissions

S. Monni (✉)
Benviroc Ltd,
Espoo, Finland
e-mail: suvi.monni@benviroc.fi

S. Syri · I. Savolainen
VTT Technical Research Centre of Finland,
Espoo, Finland

R. Pipatti
Statistics Finland,
Helsinki, Finland

 Springer

from metal, pulp and paper, and mineral industries, and from oil refineries are included in the first phase of the system, 2005–2007. The majority of emissions included are derived from combustion, but some originate from the use of raw materials. National authorities in each country have allocated initial emission permits to plants covered by the system in their national allocation plans (NAPs), which were approved by the European Commission. Altogether, emissions trading in the EU25 will cover around 2,200 Tg CO_2 emissions annually from 11,500 installations. The amount of emission allowances for the EU15 is 1,740 Tg (European Commission, 2005), which corresponds to more than 40% of CO_2-equivalent emissions from the EU15 in 2002 (the share is calculated without land use, land use change, and forestry (LULUCF); no information on activities under Articles 3.3 and 3.4 was available) (EEA, 2005; Gugele, Huttunen, Ritter, & Gager, 2004). The new member states of the European Union participating in emissions trading are expected to be mainly sellers of emission allowances during the first phase.

As emissions trading involves high monetary values, appropriate verification of emissions is needed to ensure equitable trading. Thus, the monitoring guidelines for the EU emissions trading scheme (EC, 2004) also give advice on acceptable uncertainties in plants that participate in the emissions trading scheme.

Emissions trading under the Kyoto Protocol will begin in 2008. It will cover all gases of the Kyoto Protocol (CO_2, methane [CH_4], nitrous oxide [N_2O], hydrofluorocarbons [HFCs], perfluorocarbons [PFCs], and sulfur hexafluoride [SF_6]) as well as all sectors: energy, industrial processes, waste, agriculture, and LULUCF activities defined in Articles 3.3 and 3.4 of the Protocol. Article 3.3 covers afforestation, reforestation, and deforestation; Article 3.4 covers revegetation, forest management, cropland management, and grazing land management. The rules for emissions trading under the Kyoto Protocol were adopted by the first Conference of the Parties serving as the Meeting of the Parties to the Kyoto Protocol (UNFCCC, 2005). Emissions are traded among Parties, not among companies. Parties can enable companies to trade under domestic and multilateral schemes, such as the EU ETS. Parties to the Convention have an obligation to estimate and report the uncertainties in their emission estimates to the United Nations Framework Convention on Climate Change (UNFCCC), but no bounds for uncertainty in tradable emissions are given. Proposals for the treatment of uncertainties in emissions trading are presented, for example, by Gillenwater, Sussman, and Cohen (2007) and Nahorski, Horabik, & Jonas (2007), and the effect of uncertainty on the costs of emissions trading is estimated by Godal, Ermoliev, Klaassen, & Obersteiner (2003) and Bartoszczuk and Horabik (2007). Nahorski et al. (2007) present an undershooting concept, where proving compliance is required at a specified risk level and where "effective emissions" (to be used in trading) are derived based on the selected risk factor and uncertainties in emission estimates. Gillenwater et al. (2007) also present two methods for using uncertainty estimates to adjust greenhouse gas trading ratios. Godal et al. (2003) study the outcome of the carbon permits market, given the uncertain emission levels, and examine the possibility of reducing this uncertainty by investing in monitoring. They conclude that the inclusion of uncertainty in the Kyoto Protocol would increase marginal emission reduction costs.

In this paper, we present a comparison of uncertainties in different emissions trading schemes. The aim is to present the pros and cons of different schemes from two points of view: first, we consider the importance of market size for cost-effective emission reduction; second, we estimate the changes in uncertainty introduced by the inclusion of different sources in the emissions trading. We estimate uncertainty in emissions included in the EU CO_2 emissions trading scheme (2005–2007) for both the EU15 and EU25. In addition, we present uncertainty estimates for a hypothetical scheme that also covers CH_4 and N_2O emissions from the source categories included in the EU emissions trading scheme. This example is only illustrative, as it is unlikely to occur in reality. Finally, we present estimates for the Kyoto emissions trading scheme, both with and without the forest-related activities defined in Articles 3.3 and 3.4. All uncertainty estimates are based on uncertainties in national inventories.

Section 2 presents uncertainties related to different emission sources and sinks based on the relevant literature. Section 3 outlines the methods used in this study to estimate uncertainties in emissions trading. Results are given in Section 4, and a discussion and conclusions are presented in Section 5.

2 Uncertainties in Different Emissions Trading Schemes

All emission estimates contain uncertainty. Uncertainties arise because, for example, of errors in models or measurement instruments, insufficient knowledge of the emission-generating process, or the unsuitability of the emission factors used. Annex I Parties to the Convention have the obligation under the UNFCCC to estimate and report the uncertainty in their greenhouse gas emission inventories. The countries that have performed uncertainty analyses have usually ended up with uncertainty of ±5–20% (confidence interval of 95% expressed as a percentage relative to mean value) in annual greenhouse gas emission inventories without LULUCF (Gupta, Ohlstroon, & Rotenberg, 2003; Monni, Syri & Savolainen, 2004; Rypdal & Winiwarter, 2001; Winiwarter, 2007).

It is important to differentiate between uncertainties in emission estimates for single point sources (e.g., power plants) and emission inventories. Random errors in uncorrelated emission estimates for different sources partly cancel each other out, but possible systematic errors accumulate in the national inventory.

The most accurate data in greenhouse gas emission inventories concern CO_2 from fuel combustion, also included in the EU ETS (2005–2007). Uncertainty in this emission source arises from the amount of fuel used and the carbon content of the fuel. Typically, the oxidation factor is not a significant source of uncertainty because combustion is nearly complete in large installations. For commercially traded fuels, uncertainties in the emission estimates of plants are usually around ±2.5–5% for large plants and ±5–10% for small plants (EC, 2004). Uncertainties are larger for industrial by-products, for example, coke oven gas and refinery gas (plant-specific uncertainty of ±4–10%) (e.g., EC, 2004). For waste combustion, the largest uncertainty in CO_2 emissions comes from the fossil carbon content of the fuel. Uncertainty in the carbon content of waste and the share of fossil carbon may be as high as ±50% (IPCC, 2000). In inventories, uncertainty in CO_2 from waste combustion is estimated at ±10–30% (Rypdal & Winiwarter, 2001). In the monitoring guidelines for EU emissions trading (EC, 2004), uncertainty in plant-specific emissions from waste combustion is estimated to be much lower (i.e., ±5–12.5%).

Industrial processes covered in the EU ETS are also among the best-known emission sources (e.g., limestone and dolomite used in cement and lime manufacture), although their uncertainty is typically larger than that of fuel combustion. Uncertainties in these processes are usually ±5–10% (EC, 2004; IPCC, 2000), but may be as high as ±20–40%, depending on the emission estimation method (IPCC, 2000).

In an examination of a hypothetical extended EU15 emissions trading scheme, we also included CH_4 and N_2O from the emission sources covered by the EU emissions trading scheme. CH_4 and N_2O emissions from combustion are largely dependent on process conditions (e.g., temperature in the furnace), combustion technology, and fuel quality. Uncertainty in CH_4 emissions from stationary combustion is estimated to vary between ±50 and 150% (IPCC, 2000; Rypdal & Winiwarter, 2001). N_2O emissions from combustion are even more dependent on the combustion process than methane emissions and are also sensitive to nitrogen oxide (NO_x) reduction technologies. Uncertainty in N_2O emissions from combustion is estimated to vary between ±20 and 200% (Gupta et al., 2003). IPCC (2000) estimates that uncertainty in N_2O from combustion may even be an order of magnitude. Uncertainties in emissions based on plant-specific measurements would be much smaller.

The Kyoto emissions trading scheme includes some industrial sources that are not included in the extended EU15 emissions trading scheme, for example, nitric acid and adipic acid production which can be rather accurately estimated using, inter alia, continuous measurement (e.g., ±7%) (Rypdal & Winiwarter, 2001) but the uncertainty of which may be very large (up to 230%) if emission estimation is based on calculation (Rypdal & Winiwarter, 2001). The Kyoto emissions trading scheme also covers transportation and combustion in small installations, which are somewhat more uncertain than emissions covered by the EU15 emissions trading scheme. Uncertainties in HFCs, PFCs, and SF_6 from different industrial processes vary from ±5 to 100% (Gupta et al., 2003; Rypdal & Winiwarter, 2001).

CO_2, CH_4, and N_2O emissions from agriculture and waste management are often very uncertain. Emissions from, for example, landfills, enteric fermentation of animals, and agricultural soils are difficult to estimate because all these emissions are caused by complex biological processes with various changing parameters. Uncertainties in these emissions vary from, for example, ±30–50% for CH_4 from landfills to ±75–1000% for N_2O from agricultural soils (IPCC, 2000; McGettigan & Duffy, 2003; Rypdal & Winiwarter, 2001).

Land use, land use change, and forestry is also a very uncertain emission category. Its inclusion is estimated to increase uncertainty in inventories, especially if its share of total net emissions is large (Monni et al., in press). Uncertainties in models estimating the carbon budget of forests have been presented in various studies (Dufrêne et al., 2005; Heat & Smith, 2000; Monni et al., in press; Peltoniemi, Palosuo, Monni, & Mäkipää, 2006; Smith & Heat, 2001; Verbeeck, Samson, Verdonck, & Lemeur, 2006). Changes in the carbon stocks of trees are estimated to contain an uncertainty of around ±30–100% (Monni et al., in press; Salway, Murrells, Milne & Ellis, 2002; Winiwarter & Rypdal, 2001), while emissions from liming are estimated to contain an uncertainty of ±20% (McGettigan & Duffy, 2003). Carbon stock changes in soils are estimated to be more uncertain (Ogle, Breidt, Eve, & Paustian, 2003; Paul, Polglase, & Richards, 2003a; 2003b; Peltoniemi et al., 2006; Vandenbygaart, Gregorich, Angers, & Stoklas, 2004). In addition, uncertainties in emissions or removals from land use change are estimated to be large. According to Articles 3.3 and 3.4 of the Kyoto Protocol, only a small share of the removals due to LULUCF activities can be credited in the accounting. Uncertainties in carbon stock changes from activities under Article 3.3 (afforestation, reforestation, deforestation) and Article 3.4 (forest management, revegetation, cropland management, and grazing land management) of the Kyoto Protocol are estimated to vary between ±50–100% for some activities (IPCC, 2003).

3 Materials and Methods

In this contribution, we present uncertainty estimates for five emissions trading schemes for the EU area: (1) the EU ETS (2005–2007) for the EU15; (2) the EU ETS for the EU25; (3) a hypothetical EU15 emissions trading scheme extended to cover CH_4 and N_2O; (4) EU15 emissions trading under the Kyoto Protocol without LULUCF activities; and (5) EU15 emissions trading under the Kyoto Protocol with

forest-related activities, as defined in Articles 3.3 and 3.4. Because of limitations in emissions and uncertainty data for the new EU member states, options 3–5 could not be considered for the EU25.

Emissions used in the calculation for the EU ETS for CO_2 were based on accepted national allocation plans (EEA, 2005) for 2005–2007 (Table 1). Estimated emissions of other sectors and gases were based on emissions reported in the inventory report of the EU15 (Gugele et al., 2004) and the inventory reports of the new EU member states (UNFCCC, 2004) for 2002 (Tables 1 and 2).

For the purposes of this study, emissions included in the EU ETS were divided into the following subgroups: stationary combustion (including, for example, combustion in the energy, oil refineries, pulp and paper, metal, and mineral industries); production of cement and lime (emissions from raw materials); and metal production (process emissions, for example, use of reducing agents). National allocation plans are made according to plant or activity. It is thus impossible to differentiate between emissions from combustion and processes, and the allocation of emissions to different sectors is therefore quite rough. Emissions reported under category 2.A (mineral products) of the UNFCCC were used to estimate process emissions from the mineral industry, as other emissions are due mainly to combustion. Emissions reported under category 2.C (metal production) were used to estimate process emissions from the metal industry. Emissions deriving from the use of raw materials in other industries (pulp and paper, glass, ceramics) are minor and were therefore not considered separately from emissions deriving from combustion in these sectors.

In calculating the emissions for the hypothetical extended EU15 emissions trading scheme, we assumed that CH_4 and N_2O emissions reported under the common reporting format (CRF) categories 1.A.1 (energy industries) and 1.A.2 (manufacturing industries and construction) correspond to categories of EU ETS (Table 1).

The estimate of removals from the forest sector under Articles 3.3 and 3.4 of the Kyoto Protocol was based on the estimate of maximum annual potential for carbon sequestration of forests under the first commitment period, 2008–2012, including afforestation, reforestation, and deforestation (ARD) activities, and forest management (ECCP, 2003, p. 50) (Table 2).

Thus, revegetation, cropland management, and grazing land management included in Article 3.4 of the Kyoto Protocol are not included in the estimates. Only a small number of member states has chosen to use these activities in the first commitment period (EEA, 2006).

The uncertainty estimates presented in Tables 1 and 2 were based on IPCC default uncertainties (IPCC, 2000), estimates of member states of the EU15 (Feldhusen et al., 2004; McGettigan & Duffy, 2003; Monni et al., 2004; Rypdal & Winiwarter, 2001; Salway et al., 2002), and, in the case of the EU

Table 1 Estimated emissions in energy and industrial processes sectors and corresponding uncertainties for different emissions trading schemes used as the basis for the comparisons

Area	IPCC category[1]	Emission category	Gas	Annual Emissions (Tg CO_2 eq)	Uncertainty [2]	Emissions trading scheme			
						EU ETS (EU15)	EU ETS (EU25)	Extended EU ETS	Kyoto ET
EU15	1A	Stationary combustion in large installations[3]	CO_2	1610	±3%	x	x	x	x
New EU member states	1A	Stationary combustion in large installations	CO_2	430	±7%		x		
EU15	1A	Stationary combustion in large installations	CH_4	3	±50%			x	x
EU15	1A	Stationary combustion in large installations	N_2O	20	-100 to +550%			x	x
EU15	1A	Stationary combustion in small installations[4]	CO_2	750	±7%				x
EU15	1A	Stationary combustion in small installations	CH_4	8	±50%				x
EU15	1A	Stationary combustion in small installations	N_2O	10	-100 to +550%				x
EU15	1A3	Transportation	CO_2	840	±5%				x
EU15	1A3	Transportation	CH_4	3	±50%				x
EU15	1A3	Transportation	N_2O	30	-100 to +550%				x
EU15	1B	Fugitive emissions from fuels	CO_2, CH_4	70	±30%				x
EU15	2A	Production of cement and lime	CO_2	110	±7%	x	x	x	x
New EU member states	2A	Production of cement and lime	CO_2	20	±10%		x		
EU15	2B	Chemical products	CO_2	10	±20%				x
EU15	2B	Chemical products (e.g., adipic acid and nitric acid production)	N_2O	40	±15%				x
EU15	2C	Metal industry	CO_2	20	±6%	x	x	x	x
New EU member states	2C	Metal industry	CO_2	4	±8%		x		

[1] Definitions of the categories are not exactly in accordance with definitions of the Intergovernmental Panel on Climate Change (IPCC) as they are divided between the categories included in and excluded from the EU ETS.

[2] Lower and upper bounds of 95% confidence interval expressed as percentage relative to the mean value. Symmetrical uncertainties are assumed to be as normally distributed, asymmetrical uncertainties as lognormally distributed.

[3] Plants included in EU ETS (2005–2007).

[4] Plants not included in EU ETS (2005–2007).

⌂ Springer

Table 2 Estimated emissions and corresponding uncertainties for emissions included in the Kyoto emissions trading scheme for the EU15 in addition to emissions presented in Table 1

IPCC category	Emission category	Gas	Annual Emissions (Tg CO_2 eq)	Uncertainty [1]
2	HFC emissions	HFCs	50	±40%
2	PFC emissions	PFCs	5	±40%
2	SF_6 emissions	SF_6	9	±30%
3	Solvent and other product use	CO_2, N_2O	8	±30%
4A	Enteric fermentation	CH_4	140	±40%
4B	Manure management	CH_4	70	±40%
4B	Manure management	N_2O	20	-70 to +150%
4C	Rice cultivation	CH_4	2	-80 to +200%
4D	Agricultural soils	N_2O	190	-100 to +1000%
6A	Solid waste disposal on land	CH_4	80	±45%
6B	Wastewater management	CH_4	7	±50
6B	Wastewater management	N_2O	9	-70 to +150%
6C	Waste incineration	CO_2	5	±20%
5	Forest related activities under Articles 3.3 and 3.4 of the Kyoto Protocol	CO_2	-30	±90%

[1] Lower and upper bounds of 95% confidence interval expressed as percentage relative to the mean value. Symmetrical uncertainties are assumed to be as normally distributed, asymmetrical uncertainties as lognormally distributed except N_2O from agricultural soils, which is assumed to be gamma-distributed because of high asymmetry.

ETS, on the monitoring guidelines (EC, 2004). To estimate uncertainty in activity data in the new EU countries we used the study of Suutari et al. (2001), and for the forest-related activities under Articles 3.3 and 3.4 of the Kyoto Protocol, the default estimates of the IPCC (2003) were used. All uncertainties were addressed for the EU as a whole. If it is assumed that each member state of the EU had the same relative uncertainty for a single emission category used here and that the emission estimates were independent, then this approach would overestimate uncertainty. In the real world, correlations exist, for example, when the same sources (e.g., emission factors provided by the IPCC or Corinair) or the same methods are used. On the other hand, activity data is not likely to correlate when data collection is different in each country. But some exceptions may occur, for example, in agriculture, if data used in an inventory are based on questionnaires provided in relation to EU agricultural subsidies.

We excluded emission sources whose contribution to the EU inventory in 2002 was <0.05%. These are typically emission sources reported by a single member state only. Together, these emission sources represent around 0.1% of CO_2-equivalent emissions from the EU15; thus, their effect on uncertainty can be neglected.

Uncertainties were addressed for emission categories at the EU level as presented in Tables 1 and 2. Uncertainties in different sectors were then combined using Monte Carlo simulation. In Monte Carlo simulation, random numbers are taken thousands of times from the distributions of all input parameters to obtain a probability distribution of total emissions.

4 Results

According to the calculations, emissions included annually in different EU15 emissions trading schemes range from 1740 Tg CO_2 equivalents in the EU ETS (2005–2007) to 4,110 Tg for the Kyoto emissions trading scheme. Uncertainty in the EU emissions trading scheme for both the EU15 and the EU25 was ±3%. In the hypothetical extended EU15 emissions trading scheme, which includes CH_4 and N_2O in addition to CO_2, the amount of tradable emissions increased only by less than 2% (28 Tg) but the uncertainty increased to (−4 to +8%). In the Kyoto emissions trading scheme, uncertainty was −6 to +21%, and the inclusion of forest activities under Articles 3.3 and 3.4 did not notably affect uncertainty because the volume of these activities is small compared to emissions from other sources. The results for the different emissions trading schemes are presented in Fig. 1. This shows that when the amount of emissions included in the emissions trading scheme is increased by the addition of new sources and gases, the uncertainty also increases.

5 Discussion and Conclusions

The results of this study show that there can be noticeable differences among the uncertainties in various emissions trading schemes. These results can be utilized when planning future emissions trading schemes and potential monitoring and verification procedures. The differences among the uncertainties in emissions under different emissions trading schemes were estimated based on uncertainties in national greenhouse gas inventories. The estimated uncertainties in the emissions in the different schemes ranged from ±3% for the EU ETS for CO_2 to −6 to +21% for the Kyoto emissions trading scheme, including forest-related activities under Articles 3.3 and 3.4 (ARD and forest management). Participation of the new EU countries in the EU15 CO_2 emissions trading did not noticeably increase uncertainties in emissions under the scheme.

If CH_4 and N_2O in addition to CO_2 were included in the EU emissions trading scheme (with sectors remaining the same), the market volume of emissions trading would not increase much, but the uncertainties would increase noticeably. The uncertainties could possibly be reduced with plant-specific data, but this would increase the costs of monitoring and verification. CH_4 and N_2O gases can be significant for specific processes, and the costs of reducing these emissions are sometimes lower than those of reducing CO_2 emissions. More detailed consideration of the pros and cons for the whole scheme would be needed to assess the benefits of including these gases in the scheme. The hypothetical EU15 emissions trading scheme presents only one option for extending EU emissions trading, and the implementation of this option is unlikely in the near future. Other options for extending EU ETS are, for example, the inclusion of CO_2 emissions from transportation (especially aviation) in the current emissions trading scheme.

The uncertainty in emissions included in emissions trading under the Kyoto Protocol was estimated, both

with and without LULUCF (for forest-related activities defined in Articles 3.3 and 3.4) and was found to range from −6 to +21% in both cases. Compared with the existing EU ETS, inclusion of the other sectors (especially the agriculture sector) and non-CO_2 gases introduced much additional uncertainty into the system. Inclusion of forest-related activities under Articles 3.3 and 3.4 of the Kyoto Protocol did not noticeably increase these uncertainties, as the uncertainties are of the same magnitude as for the emissions from the agriculture and waste sectors. In addition, removals under Articles 3.3 and 3.4 are estimated to be relatively small during the first commitment period, 2008–2012; thus, the inclusion of this category did not greatly affect the estimated uncertainties.

The estimate included only forest-related activities under Articles 3.3 and 3.4 of the Kyoto Protocol (the coverage was practically the same as in the forthcoming Kyoto emissions trading scheme). Not all categories or pools included in the IPCC good practice guidance for LULUCF (IPCC, 2003) were included in the estimates above; for example, the carbon stock changes in dead organic matter pools and N_2O emissions from forest soils were excluded. Emissions or removals from land use change and forestry contain some poorly understood processes that have large natural variability, and it is very difficult to

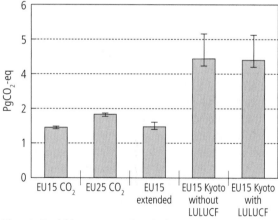

Fig. 1 Tradable amount of emissions per year and their estimated uncertainties (95% confidence interval) in different emissions trading schemes. The net emissions in the Kyoto emissions trading scheme, including the forest activities defined in Articles 3.3 and 3.4 (LULUCF), are smaller than if LULUCF is excluded; the amount of tradable permits actually increases when allowances related to both emissions and removals are traded

differentiate between natural and human-induced fluxes. LULUCF categories are thus highly uncertain (Dufrêne et al., 2005; Heat & Smith, 2000; Nilsson et al., 2000; Ogle et al., 2003; Paul et al., 2003a; 2003b; Peltoniemi et al., 2006; Smith & Heat 2001; Vandenbygaart et al., 2004; Verbeeck et al., 2006; Zhang & Xu, 2003). The inclusion of all LULUCF categories and pools is likely to increase the uncertainties in inventories (Monni et al., in press), and thus also in emissions trading.

In this study, we estimated the uncertainties in plants included in the first phase of the EU ETS (2005–2007) to be smaller than in plants not included in the scheme. This is true for current national inventories, where emissions from larger plants are often more accurately assessed than those from smaller plants because of the tighter reporting requirements in national environmental regulations for large installations. In addition, in the current monitoring guidelines (EC, 2004), more accurate estimates of activity data are required for large plants than for smaller plants. If small-scale installations (<20 MW) were also to be included in emissions trading, they could be required to meet as tight emission estimation requirements as large plants, and in that case there would be no difference between uncertainties.

In the case of emissions trading covering all the gases and sectors, results are highly sensitive to the assumptions of uncertainty in N_2O emissions from combustion and agricultural soils. The sensitivity of inventory uncertainty for the uncertainty estimates of N_2O from agricultural soils is discussed, for example, by Rypdal and Winiwarter (2001). However, an assumed numerical value for this uncertainty does not affect the fact that uncertainties are largest when all sectors and gases are included.

Under the Kyoto Protocol, Parties may use the mechanisms (emissions trading, joint implementation, clean development mechanism) "supplemental to domestic action," which must constitute "a significant element" of their efforts to meet commitments. Thus, the amount of emissions traded will, in practice, be noticeably smaller than estimated here, based on annual inventories. This will reduce the uncertainty in absolute terms.

In this study, the comparison between different emissions trading schemes was based on uncertainties in corresponding national emission inventories. In emissions trading, real uncertainties arise from emission estimates of single actors (e.g., companies and countries), and the trading of emissions uses emission allowances that are exactly defined. In practice, therefore, the uncertainties are related to actors' annual emission estimates and to monitoring and verification of the emissions. However, the approach used in this contribution gives a clear picture of the differences in uncertainties among the different emissions trading schemes.

In the future, emissions trading may cover a wider range of countries than at present. Currently, uncertainties in developing countries' emission estimates are larger than those of industrial countries. This is because of uncertainties in activity data in developing countries (e.g., less well-developed statistical systems) and lack of emission factors suitable to national conditions. In addition, uncertainty in a national emission inventory is usually larger in less-developed countries because the share of uncertain emissions (e.g., those from agriculture) is large when compared with well-known emission sources (e.g., CO_2 from fossil fuel combustion). But if the same rules of accepted uncertainty in emissions eligible for trading were applied to all countries, the participation of developing countries would not necessarily increase uncertainty. If a maximum number of countries participate in emissions trading, the emission reductions will become more cost-efficient. But, if uncertainties in emission estimates in less-developed countries will remain large, actual emission reduction benefits from emissions trading may be difficult to assess.

It is important that vendors and purchasers of emissions have similar data quality. For example, OECD (1997) suggests that tradable emissions could be discounted according to uncertainty. In this scheme, emissions with larger uncertainty would have a smaller value in emissions trading. A similar idea was presented by Gillenwater et al. (2007) and Nahorski et al. (2007).

Instead of adjusting emission estimates based on uncertainties, another option would be to divide emissions trading into parts in which uncertainties are similar. For example, emission allowances originating from increasing the carbon stock of forests could be used in the purchasing country to decrease the carbon stock of forests but not to increase emissions from fossil fuel combustion (OECD, 1997). Another possibility would be to include some kind of limit for uncertainty in emissions included in emissions trading. However, before these methods can be used, the problem of how to obtain comparable uncertainty estimates from different countries should be resolved.

Acknowledgement The authors wish to thank Wilfried Winiwarter from ARC systems research and Ger Klaassen from IIASA for their valuable comments.

References

Bartoszczuk, P., & Horabik, J. (2007). Tradable permit systems: Consiering uncertainty in emission estimates.

Dufrêne, E., Davi, H., François, C., Le Maire, G., Le Dantec, V., & Granier, A. (2005). Modelling carbon and water cycles in a beech forest. Part I: Model description and uncertainty analysis on modelled NEE. *Ecological Modelling, 185*, 407–436.

EC (2004). Commission Decision of 29/01/2004 establishing guidelines for the monitoring and reporting of greenhouse gas emissions pursuant to Directive 2003/87/EC of the European Parliament and of the Council, *C(2004) 130*, European Commission, Brussels, Belgium.

ECCP (2003). ECCP working group on forest sinks, *Final Report*, 'conclusions and recommendations regarding forest related sinks & climate change mitigation.' See http://europa.eu.int/comm/environment/climat/pdf/forest_sinks_final_report.pdf.

EEA (2005). *Climate change homepage.* Emissions trading–national allocation plans,' see http://europa.eu.int/comm/environment/climat/emission_plans.htm.

EEA (2006). The European Community's initial report under the Kyoto Protocol. Report to facilitate the calculation of the assigned amount of the European Community pursuant to Article 3, paragraphs 7 and 8 of the Kyoto Protocol Submission to the UNFCCC Secretariat, EEA, Copenhagen, Denmark.

European Commission (2005). Emissions trading: Commission approves last allocation plan ending NAP marathon. Press release, IP/05/762, 20 June.

Feldhusen, K., Hammarskjöld, G., Mjureke, D., Pettersson, S., Sandberg, A., Staaf, H., et al. (2004). Sweden's national inventory report 2004. Report submitted under the United Nations framework convention on climate change, Swedish Environmental Protection Agency, Stockholm, Sweden.

Gillenwater, M., Sussman, F., & Cohen, J. (2007). Practical applications of uncertainty analysis for National Greenhouse Gas Inventories.

Godal, O., Ermoliev, Y., Klaassen, G., & Obersteiner, M. (2003). Carbon trading with imperfectly observable emissions. *Environmental and Resource Economics, 25*, 151–169.

Gugele, B., Huttunen, K., Ritter, M., & Gager, M. (2004). Annual European Community Greenhouse Gas Inventory 1990–2002 and Inventory Report 2004, Submission to the UNFCCC Secretariat, European Commission, DG Environment, European Environment Agency, Copenhagen, Denmark.

Gupta, J., Ohlstroon, X., & Rotenberg, E. (2003). The role of scientific uncertainty in compliance with the Kyoto Protocol to the climate change convention. *Environmental Science and Policy, 6*, 475–486.

Heat, S., & Smith, J. (2000). An assessment of uncertainty in forest carbon budget projections. *Environmental Science and Policy, 3*, 73–82.

IPCC (2000). Good practice guidance and uncertainty management in national greenhouse gas inventories. J. Penman, D. Kruger, I. Galbally, T. Hiraishi, B. Nyenzi, S. Emmanuel, L. Buendia, R. Hoppaus, T. Martinsen, J. Meijer, K. Miwa, & K. Tanabe (Eds.), *Intergovernmental Panel on Climate Change*. Geneva, Switzerland.

IPCC (2003). Good practice guidance for land use, land-use change and forestry. J. Penman, M. Gytarsky, T. Hiraishi, T. Krug, D. Kruger, R. Pipatti, L. Buendia, K. Miwa, T. Ngara, K. Tanabe, & F. Wagner (Eds.), Institute for Global Environmental Strategies (IGES), Kanagawa, Japan.

McGettigan, M., & Duffy, P. (2003). Ireland, National Inventory Report. Greenhouse Gas Emissions 1990–2001. Report to the United Nations Framework Convention on Climate Change. Environmental Protection Agency, Johnstown Castle Estate, Wexford, Ireland.

Monni, S., Peltoniemi, M., Palosuo, T., Lehtonen, A., Mäkipää, A., & Savolainen, I. (in press). Uncertainty of forest carbon stock changes – Implications for the total uncertainty of GHG Inventory of Finland. *Climatic Change*.

Monni, S., Syri, S., & Savolainen, I. (2004). Uncertainties in the finnish greenhouse gas emission inventory. *Environmental Science and Policy, 7*, 78–98.

Nahorski, Z., Horabik, J., & Jonas, M. (2007). Compliance and emissions trading under the Kyoto Protocol: Rules for uncertain inventories.

Nilsson, S., Shvidenko, A., Stolbovoi, V., Gluck, M., Jonas, M., & Obersteiner, M. (2000). Full carbon account for Russia. *Interim Report IR-00-021*, International Institute for Applied Systems Analysis (IIASA), Laxenburg, Austria.

OECD (1997). Questions and answers on emissions trading among Annex I Parties. *Information Paper*, Organisation for Economic Co-operation and Development and International Energy Agency, Paris, France.

Official Journal of the European Union (2003). Common Position (EC) No 28/2003 of 18 March 2003 adopted by the Council, acting in accordance with the procedure in Article 251 of the Treaty establishing the European Community, with a view to adopting a directive of the European Parliament and of the Council establishing a scheme for greenhouse gas emission allowance trading within the Community and amending Council Directive 96/61/EC,' *Official Journal of the European Union* C 125 E/72–95.

Ogle, S., Breidt, J., Eve, M., & Paustian, K. (2003). Uncertainty in estimating land use and management impacts on soil organic carbon storage for US agricultural lands between 1982 and 1997. *Global Change Biology, 9*, 1521–1542.

Paul, K., Polglase, P., & Richards, G. (2003a). Sensitivity analysis of predicted change in soil carbon following afforestation. *Ecological Modelling, 164*, 137–152.

Paul, K., Polglase, P., & Richards, G. (2003b). Predicted change in soil carbon following afforestation or reforestation, and analysis of controlling factors by linking a C Accounting Model (CAMFor) to Models of Forest Growth (3PG), Litter Decomposition (GENDEC) and Soil C Turnover (RothC). *Forest Ecology and Management, 177*, 485–501.

Peltoniemi, M., Palosuo, T., Monni, S., & Mäkipää, R. (2006). Factors affecting the uncertainty of sinks and stocks of carbon in finnish forests soils and vegetation. *Forest Ecology and Management, 232*, 75–85.

Rypdal, K., & Winiwarter, W. (2001). Uncertainties in greenhouse gas emission inventories – evaluation, comparability and implications. *Environmental Science and Policy, 4*, 107–116.

Salway, A., Murrells, T., Milne, R., & Ellis, S. (2002). UK greenhouse gas inventory 1990 to 2000, Annual report for submission under the Framework Convention on Climate Change, AEA Technology, Harwell, UK.

Smith, J., & Heat, L. (2001). Identifying influences on model uncertainty: An application using a forest carbon budget model. *Environmental Management, 27*(2), 253–567.

Suutari, R., Amann, M., Cofala, J., Klimont, Z., Posch, M., & Schöpp, W. (2001). From economic activities to ecosystem protection in Europe. An uncertainty analysis of two scenarios of the RAINS Integrated Assessment Model,' *CIAM/CCE Report* 1/2001, IIASA, Laxenburg, Austria.

UNFCCC (2004). National Inventory Submissions 2004, Annex I. See, http://unfccc.int/ national_reports/annex_i_ghg_inventories/national_inventories_submissions/items/3473.php.

UNFCCC (2005). Modalities, rules and guidelines for emissions trading under Article 17 of the Kyoto Protocol. *Decision 11/CMP.1, CCC/KP/CMP/2005/8/Add.2*.

Vandenbygaart, A., Gregorich, E., Angers, D., & Stoklas, U. (2004). Uncertainty analysis of soil organic carbon stock change in Canadian cropland from 1991 to 2001. *Global Change Biology, 10*, 983–994.

Verbeeck, H., Samson R., Verdonck, F., & Lemeur, R. (2006). Parameter sensitivity and uncertainty of the forest carbon flux model FORUG: A Monte Carlo analysis. *Tree Physiology, 26*, 807–817.

Winiwarter, W., & Rypdal, K. (2001). Assessing the uncertainty associated with national greenhouse gas emission inventories: A case for Austria. *Atmospheric Environment, 35*, 5426–5440.

Winiwarter, W. (2007). National Greenhouse Gas Inventories: Understanding Uncertainties versus Potential for Improving Reliability.

Zhang, X.-Q., & Xu, D. (2003). Potential carbon sequestration in China's forests. *Environmental Science and Policy, 6*, 421–432.

Fig. 1 Full compliance (**a**) and the compliance with risk α (**b**) in the interval uncertainty approach

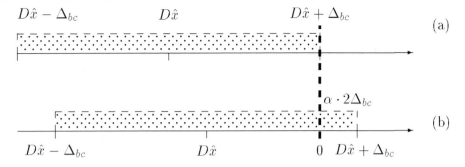

& Rypdal, 2001) gives $\zeta \sim 0.65 \div 0.7$. Thus, the dependence of inventories is quite high. It is perhaps worth mentioning that a 5% trend uncertainty is suggested as a frequent value in Monni et al. (2004b). However, this claim too is based only on a few calculated cases.

To be fully credible, that is, to be sure that a Eq. 1 is satisfied even in the worst case, the Party should prove $D\hat{x} + \Delta_{bc} \leq 0$ (see Fig. 1). Our proposition is to allow for a chance of not satisfying the obligations. In other words, we want to take a risk not greater than α ($0 \leq \alpha \leq 0.5$) that the reduction in the commitment year t_c may not be fulfilled. We then say that the Party *proves the compliance with risk α* if $D\hat{x} + \Delta_{bc} \leq 2\alpha\Delta_{bc}$ (see Fig. 1 for the geometrical interpretation). The lower bound $\alpha = 0$ corresponds to the inclusion of one-half of the uncertainty interval (full credibility). The value $\alpha = 0.5$ corresponds to completely ignoring the uncertainty. The parameter α must be set beforehand and must be common to all market participants. After simple algebraic manipulation, we obtain from the above definition the condition

$$\hat{x}_c \leq (1 - \delta)\hat{x}_b - (1 - 2\alpha)\Delta_{bc}. \qquad (9)$$

Thus, to prove compliance with risk α, the party has to undershoot its obligation by the value $(1 - 2\alpha)\Delta_{bc}$, depending on the uncertainty measure Δ_{bc}.

Alternatively, the condition (9) can be written as $\hat{x}_c + (1 - 2\alpha)\Delta_{bc} \leq (1 - \delta)\hat{x}_b$ and interpreted as upwardly correcting the emissions estimate $\hat{x}_c$, as adopted, for example, in Gillenwater et al. (2007).

The condition (9) can be also rewritten as

$$\hat{x}_c \leq [1 - \delta - (1 - 2\alpha)R_{bc}]\hat{x}_b, \qquad (10)$$

where

$$R_{bc} = \frac{\Delta_{bc}}{\hat{x}_b}$$

is the relative uncertainty half-interval with respect to the reported emissions in the base year $\hat{x}_b$. It is seen from Eq. 10 that compliance with the risk α induces a redefinition of the reduction fraction

$$\delta \quad \rightarrow \quad \delta_{Ui} = \delta + (1 - 2\alpha)R_{bc}. \qquad (11)$$

Analogously to the definition of R_{bc}, we define

$$R_b = \frac{\Delta_b}{\hat{x}_b} \qquad R_c = \frac{\Delta_c}{\hat{x}_c}$$

4 Adjustment of the Basic Committed Level

A critique of the undershooting concept could relate to the increase – of more than the agreed 5.2% – in the required reduction of reported emissions caused by the additional uncertainty-dependent expressions. This excess reduction can be corrected by shifting the reference reduction level accordingly. The idea presented here is to compare the uncertainty distributions with a reference distribution that satisfies the original obligation and has a chosen uncertainty measure. More specifically, we require the uncertainty intervals of both the reference distribution and the distribution of a Party considered to have the same upper $(1 - \alpha)$th limits. See Fig. 2 and Fig. 6. Having established this interdependency, the reduction fractions δ_{Ui} for all countries are adjusted (decreased) by the reference reduction fraction. The adjustment leaves the differences in commitment levels obtained from the undershooting but shifts them close to the original Kyoto values. In

🌍 Springer

Fig. 2 Adjustment of the committed level in the interval uncertainty approach, (**a**) reference model, (**b**) $\Delta_{bc} > \Delta_M$, (**c**) $\Delta_{bc} < \Delta_M$. $D_{Ai}\hat{x} = \hat{x}_c - (1 - \delta_{Ai})\hat{x}_b$

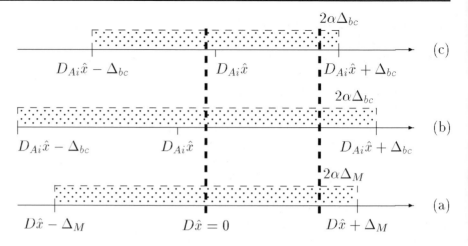

particular, the 5.2% total reported reduction is now preserved.

We assume that the center of the reference distribution exactly satisfies the committed reduction level and that its reduction fraction is therefore δ. At its upper limit of the $(1 - \alpha)$th uncertainty interval, it holds $\hat{x}_c = (1 - \delta)\hat{x}_b + (1 - 2\alpha)\Delta_M$, where Δ_M is a chosen reference half-interval. Similarly, for the same upper limit of the Party with the adjusted committed fraction δ_{Ai}, we have $\hat{x}_c = (1 - \delta_{Ai})\hat{x}_b + (1 - 2\alpha)\Delta_{bc}$. As both these upper limits have to be equal, we get the equation (see also Fig. 2),

$$(1 - \delta_{Ai})\hat{x}_b + (1 - 2\alpha)\Delta_{bc}$$
$$= (1 - \delta)\hat{x}_b + (1 - 2\alpha)\Delta_M. \quad (12)$$

This can be also written as

$$[1 - \delta_{Ai} + (1 - 2\alpha)R_{bc}]\hat{x}_b$$
$$= [1 - \delta + (1 - 2\alpha)(R_M - R_{bc}) + (1 - 2\alpha)R_{bc}]\hat{x}_b,$$

where $R_M = \Delta_M/\hat{x}_b$. This yields the following relationship for the redefinition of the reduction fraction

$$\delta \quad \rightarrow \quad \delta_{Ai} = \delta - (1 - 2\alpha)(R_M - R_{bc}). \quad (13)$$

The reduction fraction δ_{Ai} is smaller than δ_{Ui}, as the difference is

$$\delta_{Ui} - \delta_{Ai} = (1 - 2\alpha)R_M.$$

See comparison of δ_{Ai}s for different countries in Fig. 3. *Choice of R_M* An obvious choice of R_M is possibly to keep the reduction level of the Kyoto compliance unchanged. However, at least two interpretations are possible. Let us assume that N parties, $n = 1, \ldots, N$, take part in the Kyoto emission-reduction project. We can require mean committed reduction fractions before and after adjustment to be equal

$$\frac{1}{N}\Sigma_{n=1}^N \delta_A^{(n)} = \frac{1}{N}\Sigma_{n=1}^N \delta^{(n)}.$$

Inserting for $\delta_A^{(n)}$ from Eqs. 13 or 42 induces the condition

$$R_M^{av} = \frac{1}{N}\Sigma_{n=1}^N R_{bc}^{(n)} \quad (14)$$

which is the average value of all reduction fractions.

Alternatively, we can require the mean committed reduction quota to be constant:

$$\frac{1}{N}\Sigma_{n=1}^N \delta_A^{(n)}\hat{x}_b^{(n)} = \frac{1}{N}\Sigma_{n=1}^N \delta^{(n)}\hat{x}_b^{(n)}.$$

The resulting condition is a weighted average

$$R_M^{wav} = \frac{\Sigma_{n=1}^N \Delta_{bc}^{(n)}}{\Sigma_{n=1}^N \hat{x}_b^{(n)}} = \Sigma_{n=1}^N w_b^{(n)} R_{bc}^{(n)}, \quad (15)$$

where

$$w_b^{(n)} = \frac{\hat{x}_b^{(n)}}{\Sigma_{n=1}^N \hat{x}_b^{(n)}}$$

is the share of the reported emissions of Party n in the total reported emissions in the base year.

5 Uncertainties in Emissions Trading

If the above compliance-proving policy is admitted, it is possible to develop rules that include uncertainty in emissions trading and thereby solve the problem of the varying quality of this commodity among trading partners. The main line of reasoning in deriving the final formula is as follows. Assume that, during trading, the uncertainty related to the trading quota of reported emissions is transferred from the seller to the buyer. This transferred uncertainty increases the buyer's uncertainty, reducing the worth of the purchased emissions for the buyer within the compliance-proving mechanism proposed earlier. The diminished value is called an effective traded emission. It is then expressed in effective traded permits. This is the way in which the conversion ratio of reported emissions to effective permits is established. In comparison with the trading ratios between two trading partners, the effective permits form a common basis for comparison of the reported emissions for all trading.

Let us consider a selling party, recognized by the superscript S in the variables. The trend uncertainty used for proving compliance of the selling party is $\Delta_{bc}^S = (1 - \zeta)[\Delta_c^S + (1 - \delta^S)\Delta_b^S]$ or $R_{bc}^S = \Delta_{bc}^S/\hat{x}_b^S$. It then seems reasonable to assign to the sold emissions that part of the uncertainty Δ_{bc}^S that

is connected with the commitment year t_c, (i.e., $(1 - \zeta^S)\Delta_c^S$ or $(1 - \zeta^S)R_c^S = (1 - \zeta^S)\Delta_c^S/\hat{x}_c^S$). Thus, the unit $\hat{E}^S$ of the sold reported emissions brings with it the uncertainty

$$(1 - \zeta^S)\hat{E}^S R_c^S = \frac{\hat{E}^S}{\hat{x}_c^S}(1 - \zeta^S)\Delta_c^S = (1 - \zeta^S)\hat{e}^S\Delta_c^S,$$

where $\hat{e}^S = \hat{E}^S/\hat{x}_c$ is the share of the emissions units in the seller's total emissions.

If the buying country, recognized by the superscript B, purchases n units $\hat{E}^S$, then its emissions balance becomes

$$\hat{x}_c^B - n\hat{E}^S. \tag{16}$$

As countries prepare their inventories independently, it is reasonable to assume that there is *no dependence* of these estimates. Thus, we calculate the uncertainty of the buying country, after inclusion of the newly bought emissions, as

$$\Delta_{bc}^B + n(1 - \zeta^S)\hat{e}^S\Delta_c^S. \tag{17}$$

The case with dependence is discussed later. Before the trade, the following compliance-proving-with-risk-α inequality had to be satisfied

$$\hat{x}_c^B + (1 - 2\alpha)\Delta_{bc}^B \le (1 - \delta^B)\hat{x}_b^B. \tag{18}$$

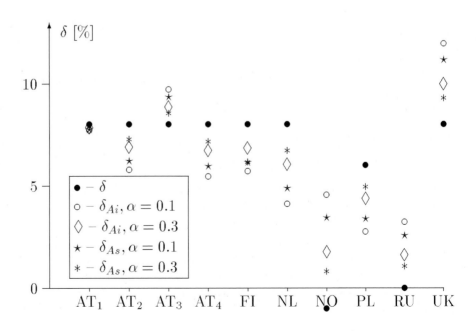

Fig. 3 Comparison of different δ_As for R_M^{av} and $\alpha = 0.1, 0.3$. δ_{Ai}-reduction fraction for the interval case, δ_{As}-reduction fraction for the stochastic case

After the trade it changes to

$$\hat{x}_c^B - n\hat{E}^S + (1-2\alpha)\left[\Delta_{bc}^B + n(1-\zeta^S)\hat{e}^S \Delta_c^S\right]$$
$$\leq (1-\delta^B)\hat{x}_b^B. \tag{19}$$

Comparing Eqs. 18 and 19 it is seen that they differ in the following component, which will be called *the effective emissions*:

$$nE_{eff} = n\hat{E}^S - n(1-2\alpha)(1-\zeta^S)\hat{e}^S \Delta_c^S$$
$$= n\left[1-(1-2\alpha)(1-\zeta^S)R_c^S\right]\hat{E}^S.$$

The effective reduction in the buyer's balance from one purchased unit $\hat{E}^S$ is

$$E_{eff} = \left[1-(1-2\alpha)(1-\zeta^S)R_c^S\right]\hat{E}^S. \tag{20}$$

Thus, the greater the seller's uncertainty, the less the purchased unit is worth to the buyer.

Note that for proving compliance, the efficient emissions are directly subtracted from the buyer's emission inventory, without any uncertainty considerations.

In the economic literature it is common to express the effects on trading by the trading ratios, see Gillenwater et al. (2007). This can easily be calculated using the effective emissions. Let $\hat{x}^B$ be the buyer's and $\hat{x}^S$ the seller's reported emissions, both equivalent to the same effective emissions x_{eff}. Thus, we have

$$\hat{x}^B[1-(1-2\alpha)(1-\zeta^B)R^B]$$
$$= x_{eff} = \hat{x}^S[1-(1-2\alpha)(1-\zeta^S)R^S].$$

The trading ratio r_t between these two parties then is:

$$r_t = \frac{\hat{x}^B}{\hat{x}^S} = \frac{1-(1-2\alpha)(1-\zeta^S)R^S}{1-(1-2\alpha)(1-\zeta^B)R^B}.$$

In the present situation the uncertainties of the prospective buyer's inventories are typically smaller than those of the prospective seller's, that is, $R^B < R^S$. In this case $r_t < 1$. This means that to allow the buyer's reported emissions to be increased by one unit, the seller must reduce $1/r_t > 1$ units of his reported emissions. By doing so, the total number of reported emissions units is reduced. Moreover, the reduction of more inaccurate reported emissions decreases the final total relative uncertainty.

Let us now discuss what the impact would be of the *dependence* of the buyer's and seller's inventories on the result. Then, according to the simplified version of Eq. 7, Eq. 17 is changed to:

$$(1-\zeta^{BS})\left(\Delta_{bc}^B + n(1-\zeta^S)\hat{e}^S \Delta_c^S\right),$$

where ζ^{BS} is the dependence parameter. Then Eq. 19 becomes

$$\hat{x}_c^B - n\hat{E}^S + (1-2\alpha)(1-\zeta^{BS})[\Delta_{bc}^B + n(1-\zeta^S)\hat{e}^S \Delta_c^S]$$
$$\leq (1-\delta^B)\hat{x}_b^B$$

causing the effective reduction of the buyer's balance

$$E_{eff} = \left[1-(1-2\alpha)(1-\zeta^{BS}v)(1-\zeta^S)R_c^S\right]\hat{E}^S \tag{21}$$

by

$$v = \frac{1}{\eta p}\frac{R_{bc}^B}{R_c^S}, \tag{22}$$

where $\eta = \frac{\hat{x}_c^B}{\hat{x}_b^B}$ is the estimated buyer's emissions reduction and $p = \frac{n\hat{E}^S}{\hat{x}_c^B}$ is the buyer's ratio of purchased emissions to total emissions in the commitment time. In Eq. 22 both η and the ratio $\frac{R_{bc}^B}{R_c^S}$ are close to 1, while p is of the order of few one-hundredths. Thus, v is big, say $50 \div 100$, and then $1-\zeta^{BS}v$ is positive only when ζ^{BS} is small enough. The existence of positive dependence on the part of the trading countries can then lead to problems in defining the effective reduction.

6 Tradable Permits Under Uncertainty

Tradable emissions permits are the usual instruments applied to limit the emissions of a pollutant. The theory of tradable permits has been elaborated for exactly known emissions Montgomery (1972). Where there are great uncertainties, as in the GHG case, our proposition is to use for permits the efficient emissions introduced in the previous section.

The effective tradable permit E_{eff} corresponding to one unit of the reported emissions $\hat{E}$ is then defined as

$$E_{eff} = \hat{E}[1-(1-2\alpha)(1-\zeta)R], \tag{23}$$

where R is the relative uncertainty of $\hat{x}$. Conversely, the reported emissions $\hat{x}$ are equivalent to $\hat{x}[1 - (1 - 2\alpha)(1 - \zeta)R]$ units of the effective tradable permits. The formula directly reflects the following rule: the higher the uncertainty, the fewer units of effective emissions permits allocated to a Party.

6.1 Compliance with Undershooting

Let us consider a Party to the Kyoto Protocol. Depending on conditions (9), in the commitment year, the Party has permission to emit $\hat{x}_c$ units of GHG, satisfying

$$\hat{x}_c \leq (1 - \delta)\hat{x}_b - (1 - 2\alpha)\Delta_{bc} =$$
$$= (1 - \delta)[1 - (1 - 2\alpha)(1 - \zeta)R_b]\hat{x}_b$$
$$- (1 - 2\alpha)(1 - \zeta)\Delta_c.$$

Adding to both sides $(1 - 2\alpha)(1 - \zeta)\Delta_c.$ and denoting, according to Eq. 23,

$$l_c = [1 - (1 - 2\alpha)(1 - \zeta)R_c]\hat{x}_c$$
$$l_b = [1 - (1 - 2\alpha)(1 - \zeta)R_b]\hat{x}_b, \qquad (24)$$

that is, the number of effective permits equivalent to the emissions $\hat{x}_c$ and $\hat{x}_b$, respectively, yields

$$\frac{1 + (1 - 2\alpha)(1 - \zeta)R_c}{1 - (1 - 2\alpha)(1 - \zeta)R_c}l_c \leq (1 - \delta)l_b.$$

As, typically, the relative uncertainty R_c for a party may be of the order $0.1 \div 0.2$, then approximately

$$(1 - \delta)\frac{1 - (1 - 2\alpha)(1 - \zeta)R_c}{1 + (1 - 2\alpha)(1 - \zeta)R_c}$$
$$\approx 1 - \delta - 2(1 - 2\alpha)(1 - \zeta)R_c.$$

We can thus use the approximation

$$l_c \leq [1 - \delta - 2(1 - 2\alpha)(1 - \zeta)R_c]l_b. \qquad (25)$$

Relation (25) expresses the commitment condition in the effective tradable permits. It has the same form as the original commitment condition for the reported emissions (9). But now, the following redefinition of the reduction fraction applies:

$$\delta \quad \rightarrow \quad \delta_{pUi} = \delta + 2(1 - 2\alpha)(1 - \zeta)R_c. \qquad (26)$$

6.2 Compliance with Adjustment of the Commitment Level

To introduce the adjustment of the basic committed level of Section 4, let us again consider the basic Eq. 12 with the new adjustment fraction δ_{pAi}

$$(1 - \delta_{pAi})\hat{x}_b + (1 - 2\alpha)(1 - \zeta)[\Delta_c + (1 - \delta_{pAi})\Delta_b]$$
$$= (1 - \delta)\hat{x}_b + (1 - 2\alpha)\Delta_M.$$

This can be written as

$$(1 - \delta_{pAi})[1 + (1 - 2\alpha)(1 - \zeta)R_b]\hat{x}_b$$
$$= (1 - \delta)[1 - (1 - 2\alpha)(1 - \zeta)R_b]\hat{x}_b +$$
$$+ (1 - 2\alpha)(1 - \zeta)[(1 - \delta)R_b\hat{x}_b - R_c\hat{x}_c]$$
$$+ (1 - 2\alpha)R_M\hat{x}_b,$$

or, using definition of l_b (24),

$$(1 - \delta_{pAi})\frac{1 + (1 - 2\alpha)(1 - \zeta)R_b}{1 - (1 - 2\alpha)(1 - \zeta)R_b}l_b =$$
$$= (1 - \delta)l_b + (1 - 2\alpha)R_M\hat{x}_b$$
$$+ (1 - 2\alpha)(1 - \zeta)\left[(1 - \delta)R_b - R_c\frac{\hat{x}_c}{\hat{x}_b}\right]\hat{x}_b.$$

Now, after similar approximate reasoning, as in the undershooting case, the above equality can be transformed as follows

$$(1 - \delta_{pAi})l_b \approx [1 - \delta - 2(1 - 2\alpha)(1 - \zeta)R_b]l_b +$$
$$+ \frac{1 - (1 - 2\alpha)R_M}{1 + (1 - 2\alpha)(1 - \zeta)R_b}l_b$$
$$+ \frac{(1 - 2\alpha)(1 - \zeta)\left[(1 - \delta)R_b - R_c\frac{\hat{x}_c}{\hat{x}_b}\right]}{1 + (1 - 2\alpha)(1 - \zeta)R_b}l_b$$

or

$$(1 - \delta_{pAi})l_b \approx [1 - \delta - 2(1 - 2\alpha)(1 - \zeta)R_b]l_b +$$
$$+ \frac{1 - (1 - 2\alpha)R_M}{1 + (1 - 2\alpha)(1 - \zeta)R_b}$$
$$\times \left(1 + \frac{(1 - 2\alpha)(1 - \zeta)\left[(1 - \delta)R_b - R_c\frac{\hat{x}_c}{\hat{x}_b}\right]}{1 - (1 - 2\alpha)R_M}\right)l_b.$$

As the following approximations can be used

$$\frac{(1 - 2\alpha)R_M}{1 + (1 - 2\alpha)(1 - \zeta)R_b} \approx (1 - 2\alpha)(1 - \zeta)R_M$$

and

$$\frac{(1-2\alpha)(1-\zeta)\left[(1-\delta)R_b - R_c\frac{\hat{x}_c}{\hat{x}_b}\right]}{1-(1-2\alpha)R_M} \ll 1.$$

Then, finally, we get approximately:

$$(1-\delta_{pAi})l_b \approx \left(1-\delta+(1-2\alpha)[R_M-2(1-\zeta)R_b]\right)l_b.$$

This provides the reduction fraction for permits with adjustment

$$\delta \quad \rightarrow \quad \delta_{pAi}=\delta-(1-2\alpha)[R_M-2(1-\zeta)R_b]. \quad (27)$$

Above, Eq. 14 or Eq. 15 can be substituted for R_M, with R_{bc} given by Eq. 8.

Calculating, as before, the difference

$$\delta_{pUi}-\delta_{pAi}=(1-2\alpha)R_M+2(1-2\alpha)(1-\zeta)(R_c-R_b)$$

we see that it is close to $\delta_{Ui}-\delta_{Ai}$, and even equal to it, when $R_c=R_b$, and is therefore, in most cases, positive.

6.3 Compliance Proving and Trading Mechanism

Thus, compliance proving and the trading mechanism with uncertain observations and adjustment of the basic committed level require the following steps.

(1) In (successive) base years, the allotted reported emissions are converted to effective permits according to the expression

$$l_b = \hat{x}_b[1 - (1 - 2\alpha)(1 - \zeta)R_b]. \quad (28)$$

(2) The committed obligations, in effective permits, in the commitment year are calculated from the condition

$$l_c \le (1 - \delta_{pAi})l_b$$
$$= \left(1-\delta+(1-2\alpha)[R_M-2(1-\zeta)R_b]\right)l_b, \quad (29)$$

which is equivalent to the reported emissions

$$\hat{x}_c = \frac{l_c}{1 - (1 - 2\alpha)(1 - \zeta)R_c}$$
$$\approx l_c[1 + (1 - 2\alpha)(1 - \zeta)R_c]. \quad (30)$$

(3) The effective permits l_c can be traded and directly added to the effective permits of any Party.

Note that if $R_M = R_{bc}$ (i.e., uncertainty of the Party equals the reference level), and $R_{bc} = 2(1-\zeta)R_b$, then $R_M = 2R_b$ and therefore $\delta_{pAi} = \delta$. In this case (29) reduces to the condition $l_c \le (1-\delta)l_b$, where the reduction fraction is equal to the original one.

The above scheme reduces trade in uncertain cases to the classic tradable permits problem. Once the reported emissions are recalculated to the effective permits, they are traded and counted for compliance proving without further consideration of the uncertainties in the emission inventories.

7 Simulation of a Carbon Market with Effective Permits

The aim of this section is to use the ideas introduced earlier in a market optimization problem (i.e., to simulate trading with effective permits within both the undershooting and adjustment framework). In constructing the market model the basic decision of each participating country is considered. Is it cheaper to abate the emissions or to buy permits on the market? The answer depends on the market price of the permit resulting from the optimization of the total cost of all participants.

7.1 Database

Before performing a carbon market simulation, the cost functions of GHG abatement need to be known for market participants. The lack of availability of data forced us to consider the original Parties to the Kyoto Protocol aggregated into five groups: United States (US), the Organisation for Economic Co-operation and Development, Europe (OECDE), Japan, Canada/Australia/New Zealand (CANZ), and finally Eastern Europe/former Soviet Union (EEFSU), instead of continuing to make calculations for the countries mentioned earlier in the paper. Data for regional abatement cost functions come from Godal and Klaassen (2003).[1] Data on

---

[1] Provision of data from Odd Godal is gratefully acknowledged.

Table 2 Base-year emissions, committed changes in emissions, inventory uncertainty, total and marginal costs of compliance without trading

	Base-year emissions	Kyoto target	Inventory uncertainty	Total costs	Marginal costs
Variable	x_i^0	σ_i	R_i	$c_i(x_i^0(1-\sigma_i))$	$\frac{\partial c_i(x_i)}{\partial x_i}$
Units	MtC/year	%	%	MUS\$	\$/tC
US	1,345	7.0	13	89,343	−313.7
OECDE	934	7.9	10	28,652	−322.7
Japan	274	6.0	15	21,077	−453.8
CANZ	217	0.7	20	10,477	−216.5
EEFSU	1,337	1.7	30	0	0.0
Total	4,107			149,549	

uncertainty levels were derived from Godal et al. (2003) and Rypdal and Winiwarter (2001) and are partly assumed for Japan. The results here, and particularly in the sequel, should be regarded as illustrative and not the ultimate solution, as the data is only partly estimated. Table 2 depicts the situation of the groups before any exchange of permits takes place, and according to the current regulations (i.e., without undershooting).

One can immediately spot from Table 2 a disproportionate gap between the Kyoto targets and the magnitude of inventory uncertainties. Although some objections can be raised about the accuracy of the uncertainty levels accepted here, the situation generally follows earlier observations (e.g., Rypdal & Winiwarter, 2001), revealing potential problems with Kyoto Protocol compliance.

7.2 No Uncertainty Market

The following notation will be used:

$n = 1, 2, \ldots, N$ – The index of a Party to the Kyoto Protocol;

$x_c^{(n)}$ – Emission level of the Party n in the commitment year;

$c^{(n)}\left(x_c^{(n)}\right)$ – Cost of reducing emissions to the level $x_c^{(n)}$;

$\delta^{(n)}$ – Fraction of the Party n base-year emissions to be reduced according to the Kyoto obligation;

$x_b^{(n)}$ – Base-year emissions of the Party n.

The task is to meet the targets of the Kyoto Protocol and not to allow the costs to become higher than necessary (Baumol & Oates, 1998; Tietenberg, 1985):

$$\min_{x_c^{(n)}} \sum_n c^{(n)}\left(x_c^{(n)}\right)$$

$$\text{s.t.} \sum_n \left(x_c^{(n)} - (1 - \delta^{(n)})x_b^{(n)}\right) = 0. \tag{31}$$

The border condition takes the form of an equation, as we assume that Parties never overcomply. Constructing the Lagrangian we obtain the condition for the static market equilibrium:

$$\lambda = -\frac{\partial c^{(n)}\left(x_c^{(n)}\right)}{\partial x_c^{(n)}},$$

where λ is the Lagrange multiplier being interpreted as the market shadow (equilibrium) price.

7.3 Market with Uncertainties

7.3.1 Effective Emission Permits

Based on the formula (23) the relationship between the reported emissions level $x^{(n)}$ and the effective emissions permits $l^{(n)}$ is

$$l^{(n)} = [1 - (1 - 2\alpha)(1 - \zeta)R^{(n)}]x^{(n)}, \tag{32}$$

where $R^{(n)}$ is the relative uncertainty of the inventory. [2]

---

[2]Here, uncertainties for the base year and for the commitment year are assumed to be equal; the subscripts b and c are thus dropped. A consideration of uncertainty reduction would require the cost of such an action also to be included in the optimization problem (31) (compare Godal et al., 2003 and Obersteiner et al., 2000).

As an effective permit will be the standard permit used in our setting, the cost of emission abatement is expressed in terms of effective permit units

$$c^{(n)}\left(x_c^{(n)}\right) = c^{(n)}\left(\frac{l^{(n)}}{1-(1-2\alpha)(1-\zeta)R^{(n)}}\right). \quad (33)$$

The argument of the abatement cost function is thus shifted according to the Party's uncertainty level $R^{(n)}$, the dependence of the commitment, the uncertainty parameter ζ of the base year, and the assumed risk level α. Market decisions will be made on the basis of the cost function (33).

7.3.2 Market with Undershooting

Having expressed abatement costs in terms of effective permits, the next step is to apply the undershooting rule so that Parties can be awarded or penalized, respectively, for their uncertainty level. Inserting from Eq. 24 for l_b in Eq. 25, the commitment condition is now expressed as follows:

$$l^{(n)} \leq [1 - \delta^{(n)} - 2(1-2\alpha)(1-\zeta)R^{(n)}]x_b^{(n)}$$
$$\times [1 - (1-2\alpha)(1-\zeta)R^{(n)}]. \quad (34)$$

This differs from the standard border condition in Eq. 31, as the original emissions obligation under the Kyoto Protocol is decreased in line with the undershooting rule according to inventory uncertainty $R^{(n)}$ and considered risk level α. The last two terms on the right-hand side of inequality (34) correspond to effective permits in the base year.

The cost-effective fulfillment of the commitments under the Kyoto Protocol expressed in terms of effective permits is now as follows:

$$\min_{l^{(n)}} \sum_n c^{(n)}\left(\frac{l^{(n)}}{1-(1-2\alpha)(1-\zeta)R^{(n)}}\right), \quad (35)$$

subject to

$$\sum_n (l^{(n)} - [1-\delta^{(n)}-2(1-2\alpha)(1-\zeta)R^{(n)}]$$
$$\times x_b^{(n)}[1-(1-2\alpha)(1-\zeta)R^{(n)}]) = 0.$$

Constructing the Lagrangian yields the condition

$$\lambda = -\frac{\partial c^{(n)}\left(\frac{l^{(n)}}{1-(1-2\alpha)(1-\zeta)R^{(n)}}\right)}{\partial l^{(n)}}. \quad (36)$$

7.3.3 Market with Adjustments

As undershooting decreases Kyoto Protocol emission liabilities, it results in an increase in abatement costs. The adjustment turns the border condition in our optimization model into the following:

$$\min_{l^{(n)}} \sum_n c^{(n)}\left(\frac{l^{(n)}}{1-(1-2\alpha)(1-\zeta)R^{(n)}}\right), \quad (37)$$

subject to

$$\sum_n (l^{(n)} - [1-\delta^{(n)} + (1-2\alpha)(R_M - 2(1-\zeta)R^{(n)})]$$
$$\times x_b^{(n)}[1-(1-2\alpha)(1-\zeta)R^{(n)}]) = 0.$$

Results for both cases of R_M^{av} and R_M^{wav} will be analyzed.

7.4 Simulation Results

Below, we present the results of the market optimization problem as formulated in Eqs. 35 and 37.

7.4.1 Trading with Effective Permits Under Undershooting

Table 3 shows the results of trading with effective permits under undershooting for a few values of the parameter α and for an assumed value of the dependence coefficient $\zeta = 0.7$ common to all Parties.[3] The table starts with $\alpha = 0.5$, which corresponds to neglecting uncertainty. Obviously, effective permits and reported emissions are equal in this case for any Party, and we obtain the standard solution with the market shadow price 142.5 \$/tC and the total abatement cost for all parties 37,150 MUS\$, very much diminished from

---

[3] Obviously, the simulation results will depend heavily on the parameter ζ. However, we do not consider the sensitivity of the results on ζ, as this parameter depends mainly on the method of inventory calculation, and one can hardly imagine tuning this parameter in practice.

Table 3 Trading with effective permits under undershooting for different levels of risk α ($\zeta = 0.7$) – results at the equilibrium points; A: marginal cost of reported emission; B: marginal cost of effective permit

	Effective emissions permits	Reported emissions	Effective permits traded	A	B	Total costs
Units	MtC/y	MtC/y	MtC/y	\$/tC	\$/tC	MUS\$
Variable	$l^{(n)}$	$x_c^{(n)}$		$\dfrac{\partial c^{(n)}(x_c^{(n)})}{\partial x_c^{(n)}}$	$\dfrac{\partial c^{(n)}(l^{(n)})}{\partial l^{(n)}}$	$c^{(n)}(l^{(n)})$
$\alpha = 0.5$						
US	1,561.6	1,561.6	310.8	−142.5	−142.5	18,433
OECDE	959.4	959.4	99.1	−142.5	−142.5	5,602
Japan	321.1	321.1	63.5	−142.5	−142.5	2,059
CANZ	248.4	248.4	32.9	−142.5	−142.5	4,583
EEFSU	807.8	807.8	−506.3	−142.5	−142.5	6,473
Total	3,898.3	3,898.3	0			37,150
$\alpha = 0.3$						
US	1,442.9	1,465.8	252.9	−195.3	−198.4	34,618
OECDE	918.7	929.8	90.9	−196.0	−198.4	10,598
Japan	304.9	310.5	61.7	−194.8	−198.4	3,848
CANZ	219.9	225.3	19.8	−193.6	−198.4	8,461
EEFSU	748.8	776.7	−425.3	−191.2	−198.4	11,658
Total	3,635.2	3,708.1	0			69,183
$\alpha = 0.1$						
US	1,327.5	1,370.3	197.0	−247.9	−255.9	55,790
OECDE	878.6	900.2	82.8	−249.8	−255.9	17,208
Japan	289.2	299.9	59.9	−246.7	−255.9	6,169
CANZ	183.0	202.8	7.7	−243.6	−255.9	13,394
EEFSU	693.5	747.3	−347.4	−237.5	−255.9	17,976
Total	3,371.8	3,520.5	0			110,537
$\alpha = 0$						
US	1,271.1	1,322.7	169.8	−274.1	−285.3	68,222
OECDE	858.7	885.3	78.7	−276.7	−285.3	21,124
Japan	281.5	294.7	59.0	−272.4	−285.3	7,525
CANZ	180.2	191.7	2.1	−268.2	−285.3	16,229
EEFSU	667.2	733.2	−309.6	−259.6	−285.3	21,482
Total	3,258.7	3,427.6	0			134,582

the situation of no trade – 149,549 MUS\$. EEFSU is the only net seller of permits.

Setting $\alpha = 0.3$ we accept the risk of 30% that a Party's actual emissions are above the Kyoto Protocol target. This is reflected in different levels of effective permits and reported emissions. The market price (marginal cost) settled on the market of effective permits $\frac{\partial c^{(n)}(l^{(n)})}{\partial l^{(n)}}$ has increased and equals 198.4 \$/tC. However, it is worth noting that marginal costs of reported emissions for each party at the equilibrium points $\frac{\partial c^{(n)}(x^{(n)}(t_c))}{\partial x^{(n)}(t_c)}$ differ, ranging from 191 \$/tC (EEFSU) to 196 \$/tC (OECDE). This reflects different levels of inventory uncertainty (Table 2). The total abatement

cost has also increased considerably to almost 70,000 MUS\$.

The situation evolves in the same direction when the parameter α is decreased further. Generally, the smaller the risk α accepted, the lower the amount of excess saved emissions for sale. For example, when α is small, then the EEFSU group can sell fewer effective permits because of the high inventory uncertainty. At the same time OECDE, with a lower inventory uncertainty, buys fewer permits. Finally, requiring undershooting of the full uncertainty belt $\Delta_{bc}^{(n)}$, as defined in Eq. 8, we would have to accept the effective permit shadow price of 285.3 \$/tC and the sum of total abatement costs of 134,582 MUS$ (compare Fig. 4). That was why adjusted Kyoto Protocol obligations also had to be examined according to Eq. 37. Note that the abatement cost in this case is still smaller than that with no trade from Table 2.

7.4.2 Trading with Effective Permits Under Adjustment

Adjusting the commitment obligation of each Party using a reference uncertainty distribution has proved to be a practical solution. The results of trading under adjustment for both R_M^{av} and R_M^{wav} are presented for $\alpha = 0$ in Table 4. As $R_M^{wav} = 0.108$ is higher than $R_M^{av} = 0.10365$, the adjusted reduction target δ_i^{Ap} is higher in the case of *average* R_M^{av}, and participants have to make more reductions. The total reported emissions equal 3,850.5 MtC/year, as compared with 3,868.3 MtC/year under the *weighted* R_M^{wav}. The permit price on the effective permit market settles at 163.6 and 158.5 \$/tC, respectively. The total abatement costs differ by 2,720 MUS$. If parameter α is changed, the influence of uncertainty can be partially relaxed (see Fig. 4), decreasing both the marginal price λ and the cost.

To sum up, the inclusion of uncertainty in the trading scheme carries some additional cost (total abatement cost in the equilibrium point) compared with the standard system, even with the adjusted target level. This is inevitable, under the assumptions made, as the abatement cost function is increasing and convex. However, this additional cost seems to be reasonable. The increase is from 37,150 MUS$ to 41,562 MUS$ (in the case of R_M^{wav}) when full uncertainty is considered ($\alpha = 0$).

8 The Stochastic Type of Uncertainty

8.1 Compliance Proving

Let us now assume that $\hat{x}(t)$ is normally distributed with the mean $E[\hat{x}(t)] = x(t)$ and variance $var[\hat{x}(t)] = \sigma^2$, with obvious notations σ_b^2 and σ_c^2 in the years $t = t_b$ and $t = t_c$, respectively. A wider class of distributions can be considered but lies outside the scope of this paper. The variable $\hat{x}_c - (1 - \delta)\hat{x}_b$ is then normal with the mean $x_c - (1 - \delta)x_b$ and the variance

$$\sigma_{bc}^2 = (1 - \delta)^2 \sigma_b^2 - 2(1 - \delta)\rho_{bc}\sigma_b\sigma_c + \sigma_c^2, \qquad (38)$$

Table 4 Trading with effective permits according to the adjusted Kyoto obligation for R_M^{av} and R_M^{wav} ($\alpha = 0, \zeta = 0.7$) – results at the equilibrium points; A: marginal cost of reported emission; B: marginal cost of the effective permit; a – $\frac{\partial c^{(n)}(x^{(n)}(t_c))}{\partial x^{(n)}(t_c)}$; b – $\frac{\partial c^{(n)}(l^{(n)})}{\partial l^{(n)}}$

	Effective emissions permits	Reported emissions	Effective permits traded	A	B	Total costs
Units	MtC/y	MtC/y	MtC/y	\$/tC	\$/tC	MUS$
Variable	$l^{(n)}$	$x^{(n)}(t_c)$		a	b	$c^{(n)}(l^{(n)})$
$R_M^{av} = 0.10365$ ($\alpha = 0$)						
US	1,475.0	1,534.8	239.7	−157.2	−163.6	22,448
OECDE	921.9	950.4	47.9	−158.7	−163.6	6,950
Japan	303.9	318.3	54.4	−156.3	−163.6	2,476
CANZ	228.7	243.3	29.4	−153.8	−163.6	5,340
EEFSU	731.4	803.7	−371.4	−148.9	−163.6	7,068
Total	3,660.9	3,850.5	0			44,282
$R_M^{wav} = 0.108$ ($\alpha = 0$)						
US	1,483.5	1,543.7	242.7	−152.3	−158.5	21,069
OECDE	924.5	953.2	46.7	−153.8	−158.5	6,523
Japan	304.9	319.3	54.2	−151.4	−158.5	2,324
CANZ	230.7	245.4	30.6	−149.0	−158.5	5,012
EEFSU	734.1	806.7	−374.2	−144.2	−158.5	6,634
Total	3,677.7	3,868.3	0			41,562

Fig. 4 Dependence of the marginal cost λ (*left*) and the cost (*right*) on α. $\zeta = 0.7$

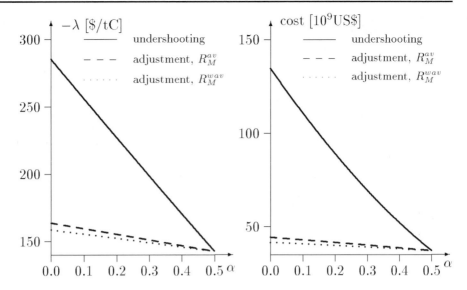

where ρ_{bc} is the correlation coefficient of $\hat{x}_b$ and $\hat{x}_c$. Calculation for a few cases provides the value $\rho_{bc} \sim 0.8$. Again, it can be seen that the correlation is high.

We require the probability of noncompliance to be not higher than α:

$$\mathcal{P}\left\{\frac{(1-\delta)\hat{x}_b - \hat{x}_c - (1-\delta)x_b + x_c}{\sigma_{bc}} \geq q_{1-\alpha}\right\} \leq \alpha,$$

where $q_{1-\alpha}$ is the $(1-\alpha)$th quantile of the standard normal distribution. This provides the condition

$$\hat{x}_c \leq (1-\delta)\hat{x}_b - (1-\delta)x_b + x_c - q_{1-\alpha}\sigma_{bc}. \quad (39)$$

If $x_c > (1-\delta)x_b$, then Eq. 39 follows from

$$\hat{x}_c \leq (1-\delta)\hat{x}_b - q_{1-\alpha}\sigma_{bc}. \quad (40)$$

If $x_c < (1-\delta)x_b$, then the committed obligation is fulfilled anyway. Thus, we conclude that fulfillment of Eq. 40 is sufficient for proving compliance

with risk α in the stochastic approach. A sketch in Fig. 5 shows the analogy in the stochastic and interval approaches.

Condition (40) can be also written as

$$\hat{x}_c \leq [1 - \delta - q_{1-\alpha}R_{bc}]\hat{x}_b,$$

where

$$R_{bc} = \frac{\sigma_{bc}}{\hat{x}_b}.$$

This case induces redefinition of the reduction fraction according to the following scheme:

$$\delta \quad \rightarrow \quad \delta_{Us} = \delta + q_{1-\alpha}R_{bc}. \quad (41)$$

Similarly we also define

$$R_b = \frac{\sigma_b}{\hat{x}_b} \qquad R_c = \frac{\sigma_c}{\hat{x}_c}.$$

A comparison of a few recalculated values of reduction commitments for the interval and stochastic case and for two values of αs are presented in

Fig. 5 Compliance with risk α in the stochastic approach

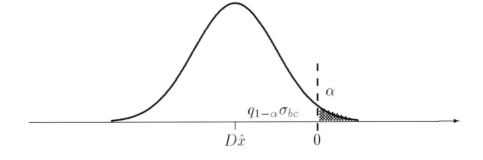

Springer

Table 5 Recalculated reduction commitments δ_{Ui} (in %)

Country	δ	R_b	R_{bc}	δ_{Ui}			
				Interval		Stochastic	
				$\alpha = 0.1$	$\alpha = 0.3$	$\alpha = 0.1$	$\alpha = 0.3$
AT	8	12	7.5	14.0	11.0	12.8	10.0
		9.8	5.1	12.1	10.0	11.3	9.3
		15	10*	16.0	12.0	14.4	10.6
		7.5	4.7*	11.8	9.9	11.0	9.2
FI	8	6	5	12.0	10.0	11.2	9.3
NL	8	4.4	3*	10.4	9.2	9.9	8.8
NO	-1	21	14.7*	10.8	4.9	8.5	2.9
PL	6	6	3.8	9.0	7.5	8.4	7.0
RU	0	17	11.9*	9.5	4.8	7.6	3.1
UK	8	19	12.8*	18.2	13.1	16.2	11.4

* Estimated using $\zeta = 0.7$.

Table 5. For those countries where R_{bc}s were not available, estimates with $\zeta = 0.7$ have been used. For $\alpha = 0.3$ and the stochastic case the shifts are not as great, even less than 1% for the smallest uncertainty and around 4% for the biggest. For $\alpha = 0.1$ and the interval case, they are much bigger, reaching almost 12% in the worst case.

8.2 Adjustment of the Basic Committed Level

Likewise, in the interval type case, for the stochastic approach we get (see Fig. 6):

$$(1 - \delta_{As})\hat{x}_b + q_{1-\alpha}\sigma_{bc} = (1 - \delta)\hat{x}_b + q_{1-\alpha}\sigma_M,$$

where σ_M is a chosen reference standard deviation. Finally

$$\delta \quad \rightarrow \quad \delta_{As} = \delta - q_{1-\alpha}(R_M - R_{bc}), \quad (42)$$

where $R_M = \sigma_M / \hat{x}_b$. See comparison of δ_{As}s for different countries in Fig. 3.

8.3 Uncertainties in Emissions Trading

In the stochastic case it is difficult to extract from σ_{bc}^S the part connected only with t_c. That is why we consider here only *uncorrelated* inventories, with $\rho_{bc}^S = 0$. It will be obvious in the sequel that this is not the only difficulty connected with the stochastic case. Thus, we admit that the unit $\hat{E}^S$

of the sold reported emissions brings with it the following uncertainty:

$$\hat{E}^S R_c^S = \frac{\hat{E}^S}{\hat{x}_c^S}\sigma_c^S = \hat{e}^S \sigma_c^S.$$

Having purchased n units, the emissions balance of the buying Party becomes

$$\hat{x}_c^B - n\hat{E}^S,$$

and its uncertainty is calculated from the expression

$$\sqrt{\left(\sigma_{bc}^B\right)^2 + \left(n\hat{e}^S\sigma_c^S\right)^2 - 2n\hat{e}^S\rho^{BS}\sigma_{bc}^B\sigma_c^S},$$

where it is assumed that a correlation exists between the trading countries' inventories, and then ρ^{BS} is the correlation coefficient of the variables $\hat{x}_c^B - (1 - \delta^B)\hat{x}_b^B$ and $\hat{x}_c^S$. To fulfill the obligations, the original emissions of the buying country should satisfy the following condition:

$$\hat{x}_c^B - (1 - \delta^B)\hat{x}_b^B + q_{1-\alpha}\sigma_{bc}^B \leq 0, \quad (43)$$

where σ_{bc} is given by Eq. 38. After purchasing $n\hat{E}^S$ units from the selling country, the new condition is:

$$\hat{x}_c^B - n\hat{E}^S$$
$$+ q_{1-\alpha}\sqrt{\left(\sigma_{bc}^B\right)^2 + \left(n\hat{e}^S\sigma_c^S\right)^2 - 2n\hat{e}^S\rho^{BS}\sigma_{bc}^B\sigma_c^S}$$
$$\leq (1 - \delta^B)\hat{x}_b^B.$$

Fig. 6 Adjustment of the committed level in the stochastic approach: **a** reference model; **b** $\sigma_{bc} > \sigma_M$. $D_{As}\hat{x} = \hat{x}_c - (1 - \delta_{As})\hat{x}_b$

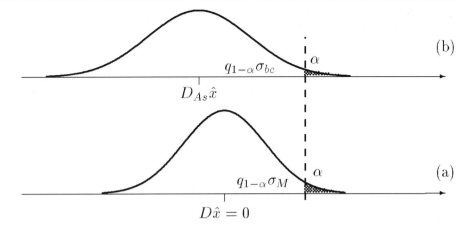

This can be written in the form

$$\hat{x}_c^B - n\hat{E}^S + q_{1-\alpha}\sigma_{bc}^B + q_{1-\alpha}$$
$$\times \left(\sqrt{\left(\sigma_{bc}^B\right)^2 + \left(n\hat{e}^S\sigma_c^S\right)^2 - 2n\hat{e}^S\rho^{BS}\sigma_{bc}^B\sigma_c^S} - \sigma_{bc}^B \right)$$
$$\le (1 - \delta^B)\hat{x}_b^B. \tag{44}$$

Subtracting Eqs. 44 and 43, and then dividing by n, we get:

$$E_{eff} = \left[1 - q_{1-\alpha}R_c^S \left(\sqrt{\left(\frac{\sigma_{bc}^B}{n\hat{E}^S R_c^S}\right)^2 + 1 - 2\rho^{BS}\frac{\sigma_{bc}^B}{n\hat{E}^S R_c^S}} \right. \right.$$
$$\left. \left. - \frac{\sigma_{bc}^B}{n\hat{E}^S R_c^S} \right) \right] \hat{E}^S. \tag{45}$$

The expression on the right-hand side is nonlinear in R_c^S, even if $\rho^{BS} \ne 0$, and cannot be reduced to a linear form similar to Eq. 20. Let us try, however, to estimate the value in the parenthesis.

Denoting the component in the parenthesis by P, it can be transformed as follows:

$$P = \frac{1 - 2\rho^{BS}v}{\sqrt{v^2 + 1 - 2\rho^{BS}v} + v},$$

where

$$v = \frac{\sigma_{bc}^B}{n\hat{E}^S R_c^S} = \frac{1}{\eta p} \frac{R_{bc}^B}{R_c^S},$$

with the same definitions as in the interval uncertainty case,

$$\eta = \frac{\hat{x}_c^B}{\hat{x}_b^B} \qquad p = \frac{n\hat{E}^S}{\hat{x}_c^B}.$$

As before, v is big, thus under the square root $1 - 2\rho^{BS}v$ can be ignored in comparison with v^2, which provides the approximate formula for E_{eff}:

$$E_{eff} = \left(1 - q_{1-\alpha}R_c^S \frac{1 - 2\rho^{BS}v}{2v} \right) \hat{E}^S.$$

Similar to the case of interval uncertainty, the value of $1 - 2\rho^{BS}v$ is positive only for the very small correlation coefficient ρ^{BS}. Thus, we assume $\rho^{BS} = 0$ to finally obtain:

$$E_{eff} = \left(1 - q_{1-\alpha}R_c^S \frac{\eta p}{2} \frac{R_c^S}{R_{bc}^B} \right) \hat{E}^S. \tag{46}$$

This formula depends not only on R_c^S, but also on the ratio $\frac{R_c^S}{R_{bc}^B}$, as well as on η and p. Due to multiplication by p, in particular, the stochastic approach gives much smaller deviations from the exact observation solutions than the interval approach. However, the dependence of both the seller and the buyer on uncertainty now makes a definition of the effective tradable permits impossible, at least in any similar way to the interval case.

9 Conclusions

The approach, presented above, of including uncertainty in reported emissions can be used to solve the problem of the different qualities of emission inventories encountered during the process of proving compliance and emissions trading caused by high and heterogeneous errors corresponding to different greenhouse gases. The advantages of the approach are its complete treatment of the uncertainty problem and that its reduction to known rules allows exact observations. In particular, the introduction of effective permits reduces the permit trade under uncertain inventory to the well-known permit trade rules with no uncertainty. To apply the approach, a knowledge of the uncertainty estimates of the inventories of all Parties involved is needed. This would stimulate research into documentation and decrease the national inventory uncertainty estimates.

Applying this approach requires different agreements among Parties participating in the emissions reduction project than is currently possible under the Kyoto Protocol. The most difficult points in negotiations might be changes in committed reductions. The proposed adjustment method makes the changes much smaller. Moreover, some free parameters may help in pinpointing the most convenient solution.

The above reasoning was centered on national emission inventories, but it can be extended to cases where uncertainties in different emitted gases are considered in trading, provided the uncertainties are not too high and justify the approximations made. The uncertainty measures R_c connected with each activity could then be used to determine the number of effective tradable permits. The idea can also be applied to other flexible mechanisms, provided the respective uncertainty measures are known for them.

While adequate conditions for undershooting and adjustment have been presented, the definition of effective permits in a stochastic case still remains unsolved because of the nonlinearities encountered. Yet, the stochastic case is important, as it better reflects reality. Moreover, for the same risk α the confidence intervals for the stochastic case are smaller than in the interval case, particularly when algebraic transformations of variables are involved, because of the effect of the concentration of probability around the mean value.

An intermediate solution can be obtained using the fuzzy uncertainty model. The calculus applied there inherits rules from the interval model, but the uncertainty may be more concentrated around the average value. An idea of this approach is mentioned in Nahorski, Jęda, Horabik and Jonas (2005) and a more mature presentation can be found in Nahorski and Horabik (2005). A generalized formula for the effective permits was thereby obtained.

The problem of the inaccuracy of the uncertainty measures, viz., the standard deviation σ or the uncertainty half interval Δ, merits closer attention. The inaccuracy of the uncertainty measure is also important from the implementation point of view. This is discussed in detail in Gillenwater et al. (2007) and will not be repeated here. The only thing perhaps worth adding is that an interim solution may be to use the uncertainty classes as proposed in Jonas and Nilsson (2007).

Another related problem is whether errors in inventories have an additive or multiplicative character (see Nahorski & Jęda, 2007). The theory in this paper assumes implicitly that the errors are additive. However, it seems to be relatively easy to adapt the results to the multiplicative errors using logarithms of the accounted data, as in Nahorski and Jęda (2007).

Acknowledgements Partial financial support from the Polish State Scientific Research Committee within the grant 3PO4G12024 is gratefully acknowledged.

References

Baumol, W. J., & Oates, W. E. (1998). *The theory of environmental policy*. Cambridge, UK: Cambridge University Press.

Charles, D., Jones, B. M. R., Salway, A. G., Eggleston, H. S., & Milne, R. (1988). *Treatment of uncertainties for national estimates of greenhouse gas emissions*. Report AEAT-2688-1. Cullham, UK: AEA Technology. See http://www.aeat.co.uk/netcen/airqual/naei/ipcc/uncertainty.

Gawin, R. (2002). *Level and trend uncertainties of Kyoto – relevant greenhouse gases in Poland*. Interim Report IR-02-045. Laxenburg, Austria: International Institute for Applied Systems Analysis (IIASA).

Gillenwater, M., Sussman, F., & Cohen, J. (2007). Practical applications of uncertainty analysis for national greenhouse gas inventories. (This issue).

Godal, O. (2000). *Simulating the carbon permit market with imperfect observations of emissions: Approaching equilibrium through sequential bilateral trade*. Interim Report IR-00-060. Laxenburg, Austria: International Institute for Applied Systems Analysis (IIASA).

Godal, O., Ermolev, Y., Klaassen, G., & Obersteiner, M. (2003). Carbon trading with imperfectly observable emissions. *Environmental and Resource Economics, 25*, 151–169.

Godal, O., & Klaassen, G. (2003). *Compliance and imperfect intertemporal carbon trading.* Working Papers in Economics No. 09/03. Bergen, Norway: Department of Economics, University of Bergen.

Gugele, B., Huttunen, K., & Ritter, M. (2005). *Annual european community greenhouse gas inventory 1990–2003 and inventory report 2005.* Technical Report No. 4/2005. Copenhagen, Denmark: European Environment Agency. http://reports.eea.europa.eu/technical_report_2005_4/en.

Gupta, J., Oltshoorn, X., & Rotenberg, E. (2003). The role of scientific uncertainty in compliance with the Kyoto protocol to the climate change convention. *Environmental Science & Policy, 6*, 475–486.

Horabik, J., & Nahorski, Z. (2004). Performance of the carbon market when accounting for uncertainties in GHG inventories. *Proceedings of the workshop uncertainty in greenhouse gas inventories: Verification, compliance & trading* (pp. 126–134). Warsaw, Poland: SRI PAS & IIASA. http://www.ibspan.waw.pl/GHGUncert2004/papers/Horabik.pdf.

Hung, W., & Wu, J. (2001). A note on the correlation of fuzzy numbers by expected interval. *International Journal of Uncertainty, Fuzziness and Knowledge-Based Systems, 9*, 517–523.

Jonas, M., Nilsson, S., Bun, R., Dachuk, V., Gusti, M., Horabik, J., et al. (2004a). *Preparatory signal detection for Annex I countries under the Kyoto protocol – A lesson for the post-Kyoto policy process.* Interim Report IR-04-024. Laxenburg, Austria: International Institute for Applied Systems Analysis (IIASA).

Jonas, M., Nilsson, S., Bun, R., Dachuk, V., Gusti, M., Horabik, J., et al. (2004b). *Preparatory signal detection for Annex I countries under the Kyoto protocol– Advanced monitoring including uncertainty.* Interim Report IR-04-029. Laxenburg, Austria: International Institute for Applied Systems Analysis (IIASA).

Jonas, M., & Nilsson, S. (2001). *The Austrian carbon database (ACDb) study – Overview.* Interim Report IR-01-064. Laxenburg, Austria: International Institute for Applied Systems Analysis (IIASA).

Jonas, M., & Nilsson, S. (2007). Prior to economic treatment of emissions and their uncertainties under the Kyoto protocol: scientific uncertainties that must be kept in mind. (This issue).

Lim, B., Boileau, P., Bonduki, Y., van Amstel, A. R., Janssen, L. H. J. M., Olivier, J. G. J., et al. (1999). Improving the quality of national greenhouse gas inventories. *Environmental Science & Policy, 2*, 335–346.

Monni, S., Syri, S., Pipatti, R., & Savolainen, I. (2004a). Comparison of uncertainty in different emission trading schemes. In *Proceedings of the workshop uncertainty in greenhouse gas inventories: Verification, compliance & trading* (pp. 106–115). Warsaw, Poland: SRI PAS & IIASA. http://www.ibspan.waw.pl/GHGUncert2004/papers/Monni.pdf.

Monni, S., Syri, S., & Savolainen, I. (2004b). Uncertainties in the finnish greenhouse gas emission inventory. *Environmental Science & Policy, 7*, 87–98.

Montgomery, W. D. (1972). Markets in licenses and efficient pollution control programs. *Journal of Economic Theory, 5*, 395–418.

Nahorski, Z., & Horabik, J. (2005). Fuzzy approximations in determining trading rules for highly uncertain emissions of pollutants. In P. Grzegorzewski, M. Krawczak & S. Zadrozny (Eds.), *Issues in Soft Computing Theory and Applications* (pp. 195–209), Warsaw, Poland: EXIT.

Nahorski, Z., Horabik, J., & Jonas, M. (2004). Greenhouse gas emission uncertainty in compliance proving and emission trading. In *Proceedings of the workshop uncertainty in greenhouse gas inventories: Verification, compliance & trading* (pp. 116–125). Warsaw, Poland: SRI PAS & IIASA. http://www.ibspan.waw.pl/GHGUncert2004/papers/Nahorski.pdf.

Nahorski, Z., & Jęda, W. (2002). Dynamics and uncertainty under Kyoto obligations. In *IIASA/FOR Workshop 'GHG Accounting: Uncertainty – Risk – Verification* (pp. 13–14). Laxenburg: International Institute for Applied Systems Analysis (IIASA).

Nahorski, Z., & Jęda, W. (2007). Processing National CO_2 inventory emission data and their total uncertainty estimates. (This issue).

Nahorski, Z., Jęda, W., Horabik, J., & Jonas, M. (2005). Propozycja zarządzania niepewnością bilansu gazów cieplarnianych w ramach protokołu z Kioto. In J. Kacprzyk, Z. Nahorski & D. Wagner (Eds.), *Zastosowania badań systemowych w nauce, technice i ekonomii* (pp. 357–372). Warsaw, Poland: EXIT.

Nahorski, Z., Jęda, W., & Jonas, M. (2003). Coping with uncertainty in verification of the Kyoto obligations. In J. Studziński, L. Drelichowski & O. Hryniewicz (Eds.), *Zastosowania informatyki i analizy systemowej w zarządzaniu* (pp. 305–317). Warsaw, Poland: SRI PAS.

Nilsson, S., Shvidenko, A., Stolbovoi, V., Gluck, M., Jonas, M., & Obersteiner, M. (2000). *Full carbon account for Russia.* Interim Report IR-00-021. Laxenburg, Austria: International Institute for Applied Systems Analysis (IIASA). (Also featured in: *New Scientist, 2253*, 18–19, August 2000.)

Nordhaus, W. D. (2005). *Life after Kyoto: Alternative approaches to global warming policies.* Working Paper 11889. Cambridge, MA: National Bureau of Economic Research. See http://www.nber.org/papers/w11889.

Obersteiner, M., Ermoliev, Y., Gluck, M., Jonas, M., Nilsson, S., & Shvidenko, A. (2000). *Avoiding a lemons market by including uncertainty in the Kyoto protocol: Same mechanism – Improved rules.* Interim Report IR-00-043. Laxenburg, Austria: International Institute for Applied Systems Analysis (IIASA).

Rypdal, K., & Winiwarter, W. (2001). Uncertainty in greenhouse gas emission inventories – Evaluation, comparability and implications. *Environmental Science & Policy, 4*, 104–116.

🍃 Springer

Rypdal, K., & Zhang, L.-C. (2000). *Uncertainties in the Norwegian greenhouse gas emission inventory*. Report 2000/13. Oslo, Norway: Statistics Norway.

Salway, A., Murrells, T., Milne, R., & Ellis, S. (2002). *UK greenhouse gas inventory, 1990 to 2000: Annual report for submission under the framework convention on climate change*. Didcot, UK: AEA Technology.

Tietenberg, T. H. (1985). *Emission trading, an exercise in reforming pollution policy*. Washington DC: Resources for the Future Inc.

van Amstel, A. R., Olivier, J. G. J., & Ruyssenaars, P. G. (Eds.) (2000). *Monitoring of greenhouse gases in the Netherlands: uncertainties and priorities for improvement*. Report 773201 003. Bilthoven, The Netherlands: National Institute of Public Health and the Environment.

Victor, D. G. (1991). Limits of market-based strategies for slowing global warming: the case of tradable permits. *Policy Sciences, 24*, 199–222.

Vreuls, H. H. J. (2004). Uncertainty analysis of Dutch greenhouse gas emission data, a first qualitative and quantitative (TIER2) analysis. In *Proceedings of the workshop uncertainty in greenhouse gas inventories: Verification, compliance & trading* (pp. 34–44). Warsaw, Poland: SRI PAS & IIASA. http://www. ibspan. waw.pl/GHGUncert2004/papers/ Vreuls.pdf.

Winiwarter, W. (2007). National greenhouse gas inventories: understanding uncertainties versus potential for improving reliability. (This issue).

Winiwarter, W., & Rypdal, K. (2001). Assessing the uncertainty associated with national greenhouse gas emission inventories: a case study for Austria. *Atmospheric Environment, 35*, 5425–5440.

Water Air Soil Pollut: Focus (2007) 7:559–571
DOI 10.1007/s11267-006-9109-3

The Impact of Uncertainty on Banking Behavior: Evidence from the US Sulfur Dioxide Emissions Allowance Trading Program

Olivier Rousse · Benoît Sévi

Received: 12 January 2007 / Accepted: 12 February 2007 / Published online: 16 February 2007
© Springer Science + Business Media B.V. 2007

Abstract In this paper, we study empirically whether uncertainty has an influence on trade in the US sulfur dioxide allowances market. In particular, we investigate the role of uncertainty on banking behavior. To do this, we introduce a tractable, structural model of trading permits under uncertainty. The model establishes a relation between banking behavior and risk preferences, especially prudence in the Kimball (1990) sense. We then test this model using data on allowances, for utilities submitted to the US Environmental Protection Agency's Acid Rain Program, carried over from one year to the next. Evidence is found of imprudence, namely, utilities bank permits in order to favor higher profits. Another finding is that larger utilities do not adopt behavior significantly different from that of smaller ones.

This paper was presented at the "International Workshop on Uncertainty in Greenhouse Gas Inventories: Verification, Compliance & Trading" in Warsaw, Poland, September 2004, under the title "Portfolio Management of Emissions Permits and Prudence Behavior."

O. Rousse (✉) · B. Sévi
Centre de Recherche en Économie et Droit de l'Énergie, Faculté des Sciences Economiques, Université Montpellier I, Av. de la Mer – Site de Richter, CS 79606, 34960 Montpellier Cedex 2, France
e-mail: olivier.rousse@univ-montp1.fr

Keywords emissions trading · permits banking · acid rain program · uncertainty · risk aversion · prudence

1 Introduction

The literature on emissions trading began with the work of Dales (1968), who introduced a number of main characteristics and critiques concerning the use of these markets as tools to control pollution. The first theoretical discussions were revived by large-scale projects and implementations of such programs. Among these programs are the American experiences (e.g., the Acid Rain Program, the Ozone Transport Commission [OTC] nitrogen oxide [NO_x] Budget Program, and the RECLAIM Program); the European emissions trading scheme that began in January 2005; and the future global greenhouse gas market under the auspices of the Kyoto Protocol (1997) of the UNFCCC.

At present, it is widely recognized that, under the hypothesis of a perfect market, a system of emissions permits is a flexible instrument to attain an environmental objective at the lowest aggregate cost. These cost savings come from averaging and trading (intrafirm and interfirm flexibility; for theoretical proofs, see, for example, Cropper & Oates, 1992; Montgomery, 1972; Tietenberg, 1985) and from banking (intertemporal flexibility;

⌲ Springer

for theoretical proofs, see, for example, Cronshaw & Kruse, 1996; Kling & Rubin, 1997; Rubin, 1996; Tietenberg, 1985). Unfortunately, perfect market assumptions rarely hold in practice. Indeed, emissions permits markets can suffer from several impediments, such as uncertainties, transaction costs (see Cason & Gangadharan, 2003; Montero, 1997), market power (see Hahn, 1984; Liski & Montero, 2005; Misiolek & Elder, 1989; van Egteren & Weber, 1996), and cheating behaviors (Keeler, 1991; Malik, 1990, 2002).

In this paper, we focus our attention on uncertainty. Large-scale experiences have shown that well-designed markets minimize transaction costs, cheating behaviors, and the risk of the exercise of market power. However, they do not succeed in reducing the various sorts of uncertainty that firms may face in such markets, including permit price uncertainty; demand uncertainty, which means production and emissions uncertainty; abatement costs uncertainty; and regulatory uncertainty. A number of researchers have analyzed the role of uncertainty in emissions permits markets. The first conclusions come from experimental economics. Carlson and Sholtz (1994) and Godby, Mestelman, Muller and Welland (1997) have shown, in different experimental settings, that uncertainty faced by regulated firms regarding their total emissions creates greater price instability than when banking is not allowed. Moreover, price peaks are higher in periods of high emissions. In a theoretical and numerical paper about marketable permits, Montero (1997) analyzes the effects of trade approval and transaction cost uncertainties on market performance and aggregate control costs. Although uncertainty and transaction costs suppress exchanges that otherwise would have been mutually beneficial, it is shown that a marketable permit system is still cost-effective compared with a command-and-control approach.

In a model of perfectly competitive markets, Hennessy and Roosen (1999) examine the impact of stochastic pollution on production decisions. They show that the existence of uncertainty as to the magnitude of pollution tends to reduce production activities – an effect à la Sandmo (1971) – compared with the situation of nonstochastic

pollution with the same mean rate of emissions.[1] Ben-David, Brookshire, Burness, McKee, and Schmidt (2000) also assume risk aversion to analyze the effects of permit price uncertainty on firms' abatement investments and trading behaviors. Experimental results suggest that abatement efforts of risk-averse permit sellers (buyers) are lower (higher) under uncertainty than under certainty. Consequently, at equilibrium, the number of allowances traded is lower under uncertainty than in a perfect market setting. Recently, Baldursson and von der Fehr (2004) obtained a similar result using the concept of risk aversion to qualify trading attitude: "when firms are sufficiently risk averse trade will be limited; in particular, infinitely risk-averse firms would not trade at all" (p. 696).

Note that the financial aspect of emissions trading is ignored throughout the literature. The majority of papers mentioned here use a static framework and do not take into account any temporal effect of price discovery. This weakness may be explained by the environmental economics approach, which does not deal with intertemporal pricing and subsequent portfolio management.

The aim of this paper is to fill a gap in the literature of emissions trading under uncertainty by providing an analytical and empirical evaluation of the banking behavior of utilities under uncertainty using the concept of prudence developed by Kimball (1990). Our methodology is similar to that used in a consumption framework where authors aim to indicate if motivation for precautionary saving is increased in response to uncertainty concerning future income. Our proxies for the uncertainty that utilities are faced with are (1) the share of coal-based generation for the utility; and (2) whether the utility is located in a deregulated or regulated state. Econometric results provide evidence that utilities respond to uncertainty by banking emissions permits, particularly when their power is mainly coal-generated. However, we do

---

[1] The authors argue that firms' behavior should be represented through a risk-averse utility function because of the natural aversion to dismissal on the part of managers (Hennessy & Roosen, 1999, p. 221).

not find a stronger motivation for banking in states where restructuring is active than in those where it is not.

The next section continues with a presentation of the sulfur dioxide (SO$_2$) emissions allowances market and a review of previous economic studies of permit banking issues that are relevant to this paper. Section 3 provides a simple model of emissions trading under uncertainty. The model gives necessary and sufficient conditions for banking given the risk preferences of the firm. Sections 4 and 5 describe data and econometric specification, respectively. Empirical estimations are discussed at the end of Section 5. Concluding remarks follow in Section 6.

2 The Sulfur Dioxide Market, Uncertainty, and Banking

The US EPA's Acid Rain Program, which began in 1995, is the first large-scale, long-term environmental program using marketable permits to tackle air pollution. The program requires utilities to reduce their emissions of sulfur dioxide (SO$_2$) by 10 million tons below 1980 levels by 2010. The program is divided into two phases. Phase I began in 1995 and affected 263 utility units at 110 mostly coal-burning electric power plants located in 21 eastern and midwestern states. An additional 182 units joined Phase I of the program as substitution or compensating units, bringing to 445 the total number of units affected during Phase I. Phase II began in 2000, tightening the annual emissions limits imposed on these large, higher-emitting plants. Phase II also set restrictions on smaller, cleaner plants fired by coal, oil, and gas, encompassing over 2,000 units in all. The program affects existing utility units serving generators with an output capacity of greater than 25 megawatts (MW) and all new utility units. Actually, every major fossil fuel-burning power production facility in the United States is now affected under Title IV of the Clean Air Act Amendment.

Each year, the US EPA distributes allowances based on a uniform national emissions rate multiplied by the utility's previous use of coal. At the end of the compliance period, a utility must hold allowances at least equal to its yearly emissions. Firms are free to trade permits and can also bank excess allowances for future use, or sell them in subsequent compliance periods. Significant penalties are applied to firms that do not comply with this rule. A brief summary of the Acid Rain Program design is given in Table 1.

Many studies have already analyzed the functioning of the US SO$_2$ allowances market, especially Phase I (see, e.g. Bohi & Burtraw, 1997; Burtraw, 1996; Ellerman & Montero, 1998; Ellerman, Joskow, Schmalensee, Montero, & Bailey, 2000; Hahn & May, 1994; Joskow & Schmalensee, 1998; Joskow, Schmalensee, & Bailey, 1998; Schmalensee, Joskow, Ellerman, Montero, & Bailey, 1998; Swift, 2001). From these studies, it appears that firms may face unexpected developments in the emissions permit market. For example, the first years of the program were characterized by low price levels compared with forecast levels. More precisely, in the beginning of the year 1996, the price of allowances fell below US$70, whereas early price estimates were in the range of US$300–1,000[2] (see Hahn & May, 1994). There are several explanations for the low price levels observed. First, the discounting of future costs led firms to make large investments in scrubbers and to bank allowances for future use. Second, the unanticipated widespread availability of low-sulfur coal due to the deregulation of railroads[3] decreased marginal costs. Third, competition from low-sulfur coal raised innovation in scrubber technologies. Fourth, forecasts could not exactly predict the general equilibrium effects caused by the emissions permits, for example, on electricity demand. Fifth, bonus allowance subsidies for scrubbers and also substitution and compensation units (the "Opt-in Program") delayed future costs. And finally, the two phases of the program segregated sellers and buyers of permits.

Generally speaking, these unanticipated developments in the allowances market show that the

---

[2] Resource Data International: US$309; American Electric Power: US$392; Sierra Club: US$446; EPRI: US$688; Ohio Coal Office: US$785; United Mine Workers: US$981.

[3] Under the Staggers Rail Act of 1980; see Ellerman and Montero (1998).

Table 1 The design of the acid rain program

Aim	Prevention of acid rain (SO_2 emissions regulation)
Duration	1995–2030
Unit value of a permit	1 ton of SO_2
Spatial coverage	United States
Sectoral coverage	Electricity-generating units (essentially coal-burning plants)
Compliance	At the firm level
Opt-in Program	Yes
Number of phases	Two (1995–1999 and 2000–2030)
Compliance period	Annual
Borrowing of permits	No
Banking of permits	Yes
Initial allocation	Free annual allocation and 3% by auction
Access for new entrants	Purchase of allowances on the market
Organizational design	Over-the-counter, or more often, via a broker
Tracking system	Allowance Tracking System (ATS)
Penalty	US$2,000/ton and reduction of permits for next year (ratio 1:1)
Access to trading	Free for every legal entity or natural person

markets for emissions permits are extremely risky. In other words, allowance prices are volatile. Figures 1 and 2 show that, as the SO_2 allowances market has matured and as prices have escalated during the past year, the long-term volatility has increased significantly. In practice, permit price uncertainty appears to be one of the main problems regulated firms face in making compliance decisions. For example, a great number of factors can suggest that permit prices may rise. Among these factors are the possibility of increases in electricity demand or fossil fuel prices, possible growth of permit demand because of new pollution sources, or a potential drastic reduction of emissions in a future phase of the program.

So, like oil, gas, coal, or electricity, emissions permits are commodities with market values that require proactive portfolio management by regulated firms, even if they are allocated free of charge. In the Acid Rain Program, the value of the emissions permits portfolio of an electricity producer often exceeds US$500 million, with market price volatility about 40–60%. Thus, when electricity producers keep all or part of their allowances in their portfolio, they take a speculative position, relying on their expectations of permit prices and electricity demand.

In this sense, pollution permits may be seen as commodities, or rather as forward contracts on commodities, that can be traded freely. The

Fig. 1 Market-clearing prices in the SO_2 market (1999–2004)

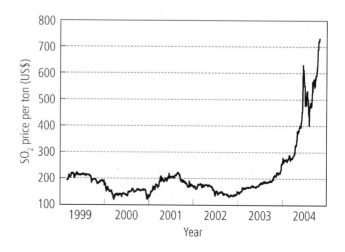

Fig. 2 Price volatility in
the SO₂ market
(1999–2004)

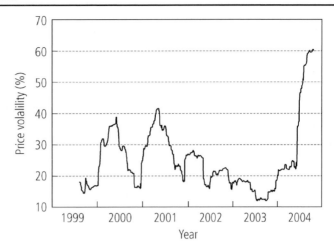

difference with standard inputs is that permits are
not needed for production to begin. Emissions
markets are designed in such a way that it is
currently possible to produce without a permit
because production periods do not coincide with
the compliance period. This is why we consider
emissions permits as forwards and not as spot
commodities.[4]

Thus, after the initial allocation of permits, regulated firms must choose whether to keep their
allowances in their portfolio or to sell them and
buy them back later. At constant prices, if a firm
sells some permits and buys them back later at a
lower price, it realizes a gain.[5] However, if this
firm sells permits and buys them back later at a
higher price, then it suffers a loss. Consequently,
a firm that is long in permits may hesitate to sell
permits if there is a chance it will need them at a
later date.

This result suggests that firms may have different banking strategies depending on their risk
exposure and risk perception. Theoretically, it
has been recognized that with perfect foresight,
permit trading, banking, and borrowing lead to
an efficient allocation of permits that collectively
minimizes cost (Rubin, 1996). In practice, the
borrowing of permits is not allowed for environ-

mental reasons; to get around this prohibition,
firms are lobbying to reduce the cap at the end
of the program. When trading in permits and
banking are allowed, the rate of change in the
price of emissions follows a simple Hotelling's rule
(Cronshaw & Kruse, 1996; Rubin, 1996). In fact,
when the permit stocks are positive and the non-
negativity constraint on permits is not binding, the
allowance price rises at the rate of interest. Using
optimal control theory, Kling and Rubin (1997)
find similar results and show that firms have incentives to bank permits when marginal abatement
costs are rising, marginal production costs are
falling, emissions standards are increasing, or output prices are rising. The only study that considers
the emissions permits market under uncertainty
is that by Schennach (2000). In her model, risk-
neutral firms minimize their expected discounted
costs. In this setting, the rate of change in the price
of emissions does not necessarily follow a simple
Hotelling's rule. Notably, when firms anticipate
that there is the possibility of a permit stockout,
the expected change in marginal abatement costs
could be negative. These permit stockout expectations could partially explain normal backwardation, that is, when prices for permits for this period
exceed those for future periods.[6]

---

[4]To understand the difference between spot and forward,
remember that a permit is always designed for a given
compliance period.

[5]Provided that transaction costs are not too high and the
interest rate is higher than inflation.

[6]Bailey (1998) provides empirical evidence of backwardation. Note that, for convenience, in our model we suppose
unbiasedness (i.e., neither backwardation nor contango).
However, our results remain valid even in a normal backwardation case.

3 A Model of Emissions Trading under Uncertainty

This section describes a simple underlying model to perform econometric estimations. Consider a competitive firm that sells a single output. The quantity $\tilde{q}$ is not known prior to the emissions trading decision. In addition, the firm faces two other sources of uncertainty, namely, the selling price per unit $\tilde{p}$ and the price of permits $\tilde{c}$ (the support for $\tilde{c}$ is $[\underline{c}, \overline{c}]$). The wealth π_0 is an initial wealth, which incorporates the initial endowment of emissions allowances at date $t = 0$. We take a different approach from that of Baldursson and von der Fehr (2004) by assuming that initial endowment has no effect on the optimal trading decision because of the opportunity cost of selling permits at the market-clearing price.[7] We assume that $\tilde{q}$ and $\tilde{c}$ are positively correlated through a simple linear relation:

$$\tilde{q} = \mu + \delta\tilde{c} + \tilde{\varepsilon}, \qquad (1)$$

where $\tilde{\varepsilon}$ is a zero-mean random variable independent of $\tilde{c}$ and δ is a positive scalar. The expected quantity is then

$$\mu + \delta E(\tilde{c}).$$

The justification for a positive relation between output quantity and permit price is intuitive (see Chicago Climate Exchange, 2004). The profit of the firm with a constant marginal cost r and a volume of permits held h is given by

$$\tilde{\pi} = \pi_0 + \tilde{q}(\tilde{p} - \tilde{c} - r) - h(c_f - \tilde{c}). \qquad (2)$$

We assume that the firm can trade only at $t = 0$. No trade is possible between $t = 0$ and $t = 1$. At $t = 1$, all uncertainties are resolved. It can be observed that, unlike previous studies, we do not take into account any abatement costs. Indeed, abatement costs have an impact on the optimal allowances trading strategy of a firm, through the now well-known property that – in the absence of

banking – marginal abatement cost should equal permit price (see Montgomery, 1972). However, at the end of 2001, permit prices were decreasing (see Fig. 2). We can then consider that new investment decisions in abatement technologies cannot be made at this period.[8]

The optimal volume of permits to hold maximizes the expected-utility profit of the firm, which is assumed to have a standard von Neumann–Morgenstern utility function ($u' > 0$ and $u'' < 0$ indicating risk aversion). The program is then

$$\max_h [Eu(\tilde{\pi})]. \qquad (3)$$

Because the second-order condition is satisfied given the concavity of the utility function, the following first-order condition is a necessary and sufficient condition for a unique maximum:

$$E[u'(\tilde{\pi})(\tilde{c} - c_f)] = 0. \qquad (4)$$

For any two random variables, $\tilde{x}$ and $\tilde{y}$,

$$E(\tilde{x}\tilde{y}) = E(\tilde{x})E(\tilde{y}) + cov[E[\tilde{x} \mid y], \tilde{y}].$$

Condition 4 can then be rewritten as

$$[c_f - E(\tilde{c})]E[u'(\tilde{\pi})] = cov[E[u'(\tilde{\pi}) \mid c], \tilde{c}]. \qquad (5)$$

If the SO_2 allowances market is unbiased (or $c_f - E(\tilde{c}) = 0$), as shown empirically by Albrecht, Verbeke, and de Clercq (2004), then optimality requires

$$cov[E[u'(\tilde{\pi}) \mid c], \tilde{c}] = 0.$$

The following proposition establishes our central result:

Proposition 1 *Consider the emissions allowances market as unbiased, then a risk-averse and prudent firm will optimally hold a volume of allowances below the corresponding level for its expected output.*

---

[7] Note that in Baldursson and von der Fehr (2004), the initial allocation of permits, investment decisions, and compliance occur simultaneously.

[8] Of course, ignoring the firms' abatement policies is not standard in emissions trading theory. Nevertheless, it does not weaken our empirical results, because of the particular period considered.

Proof The proof is by contradiction. Differentiating $E[u'(\tilde{\pi}) \mid c]$ with respect to c yields

$$\frac{\partial E[u'(\tilde{\pi}) \mid c]}{\partial c} = E[(\delta \tilde{p} - \mu - \delta r - \tilde{\varepsilon}$$

$$- 2\delta \tilde{c} + h)[u''(\tilde{\pi}) \mid c]]$$

$$= [h - E(\tilde{q}) - \delta [E(\tilde{c}) + r$$

$$- E(\tilde{p})]] E[u''(\tilde{\pi}) \mid c]$$

$$- cov[\tilde{q}, [u''(\tilde{\pi}) \mid c]].$$

If $cov[E[u'(\tilde{\pi}) \mid c], \tilde{c}] = 0$, then $\frac{\partial E[u'(\tilde{\pi})|c]}{\partial c}$ cannot be uniformly negative or positive on the support $[\underline{c}, \overline{c}]$.

First, consider the firm as prudent ($u''' > 0$). Then

$$cov[\tilde{q}, [u''(\tilde{\pi}) \mid c]] > 0,$$

because the profit $\tilde{\pi}$ is an increasing function with respect to the quantity $\tilde{q}$. It follows that $h - E(\tilde{q}) < 0$ to obtain $\frac{\partial E[u'(\tilde{\pi})|c]}{\partial c}$ not uniformly negative.

The case corresponding to $u''' < 0$ is symmetric.

□

The result appears counterintuitive at first sight. If the firm is prudent (in the sense of Kimball, 1990),[9] it should optimally hold a volume of emissions allowances below that corresponding to the expected output.[10] Inversely, an imprudent firm should hold a higher volume compared with the expected output. This ambiguous result comes from the difference between prudence à la Kimball and prudence in the everyday sense.[11] Initially, prudence emerges in a consumption setting to explain precautionary saving for an agent facing a future income risk. The aim of the

prudent agent is to smooth consumption over time. A parallel can be drawn in a production framework. In order to smooth profits, the prudent firm has an incentive to shift part of the profit from higher realizations to lower realizations.

To be more precise, because of the positive relation between quantity (electricity demand) and permit price, two cases must be considered. The first case is positive. If demand is high, profits will be increased by holding allowances, because the firm will not have to purchase additional allowances at a higher price. But inversely, in the second case, if demand is low, the firm will lose both on output sales and on allowances sales. This is because the firm will have to sell excess permits at a lower price, which is itself induced by low demand. Thus, by holding a lower volume of allowances, the utility faces no risk in losing both on output and on allowances. Nevertheless, in the positive case, the profit will be lower. The model aims to test whether such behavior exists in the SO_2 allowances market. Concretely, are utilities prudent or imprudent?

4 The Data

To obtain aggregated data at the utilities level,[12] three different information sources are needed: the EPA's Allowance Tracking System (ATS) database, the Emissions & Generation Resource Integrated Database (eGRID) 2002 database, and the Annual Electric Power Industry database of the US Department of Energy's Energy Information Administration.

The EPA is responsible for recording the transfer of allowances that are used for compliance and confirming that utilities hold at least as many

---

[9]See Gollier (2001) for a presentation of the concept of prudence.

[10]Note that if firms' preferences are assumed to be quadratic, then the separation property (Holthausen, 1979) applies and the optimal number of permits to hold corresponds to the expected output level.

[11]This difference is pointed out by Eeckhoudt and Gollier (2005). The authors consider the case of self-protection to illustrate the counterintuitive meaning of prudence in the Kimball (1990).

[12]To capture heterogeneity fully, the Arimura (2002) model examines decisions at the generating-unit level. In contrast, the analysis by Bailey (1998) is at the state level and that of Considine and Larson (2004) considers the holding level. For our study, the utilities level is the most relevant. The decisions concerning banking or trading cannot reasonably be made at the generating-unit level. Similarly, the holding level may be considered too synthetic.

🖄 Springer

Fig. 3 SO$_2$ allowances transferred under the acid rain program

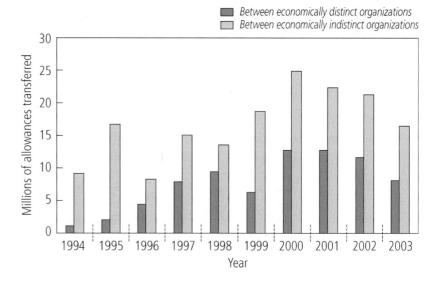

allowances as tons of SO$_2$ emitted (see Fig. 3). The ATS is the official record of allowance holdings and transfers.[13] These data are included in Appendix A of the Acid Rain Program Annual Progress Report, published on the EPA Web site.[14] For each generating unit,[15] the allowances allocated for the year, the allowances held in accounts at the end of the year, the allowances deducted at the end of the year, and the allowances carried over to the next periods are provided.[16] We then aggregate data at the plant level.

The eGRID is a comprehensive database of environmental attributes of electric power systems, prepared by the EPA Office of Atmospheric Programs and E.H. Pechan & Associates Inc. eGRID is based on available plant-specific data for all US electricity generating plants. eGRID 2002 includes nonutility power plants as well as utility-owned plants with data for 1996–2000. From 1998 on, plant-level data are available for both utility and nonutility plants. We make eGRID data coincide with EPA ATS data for each plant considered. We obtain a vector of characteristics, including the plant's generator capacity (in MW), annual net generation (in megawatt hours, MWh), annual SO$_2$ emissions (tons), annual SO$_2$ output emissions rate (pounds/MWh), annual net generation (MWh) by fuel, and other, more specific features. This vector is now related with allowances data.

Finally, the Annual Electric Power Industry database (Form EIA-861 database) contains aggregate operational data at the utilities level. These characteristics include quantitative variables such as retail revenue, resale revenue, delivery revenue, or other revenues, as well as a fundamental qualitative variable for our study, namely, ownership type.

By aggregating data at the utilities level, we obtain characteristics for about 67.86% of the total sample – in allowances volume – described in the EPA ATS database.[17] For other plants, it is not possible to determine the owner's name in the eGRID database satisfactorily. This may be because of mergers and acquisitions, or errors and gaps in the database.

---

[13]Unfortunately, the ATS does not provide any price information.

[14]http://www.epa.gov/airmarkets/progress/index.html

[15]Each plant is divided in several generating units or boilers.

[16]Of course, the number of allowances carried over to the next year can be calculated by subtracting the allowances deducted at the end of the year from the allowances held in accounts at the end of the year.

---

[17]The 137 remaining utilities are listed in the Appendix to this chapter.

5 Estimation and Empirical Findings

Our formulation is similar to formulations in consumption and saving studies, where prudence and precautionary saving are estimated (see, e.g., Skinner, 1988; Kazarosian, 1997; Lusardi, 1998; for a detailed survey, see Browning & Lusardi, 1996). The aim of these papers is to investigate whether future income risk has a significative impact on saving behavior – namely, precautionary saving – following the theoretical formulation of Kimball (1990). Our aim is identical, but in a production framework, in that we want to measure the impact on banking behavior of future uncertainty faced by utilities. Because trading is influenced by many variables, we cannot estimate a coefficient for prudence. We restrict our attention to testing for the "precautionary motive" for banking.

We now describe how banking and uncertainty will be measured for our empirical test.

5.1 Allowances Banking Behavior

For each utility, we calculate a ratio measuring the intensity of banking. Let ρ_i be the number of allowances allocated for 2001, τ_i be the number of allowances carried over to 2002, and η_i be the number of allowances deducted in 2001. The ratio is given by

$$ratio_i = \frac{(\rho_i + \tau_i) - g\eta_i}{\eta_i},$$

with g being the expected growth rate for total electricity sales in the United States. Following the *Annual Energy Review 2003* from the Energy Information Administration (2004), the expected growth rate was about 4.75% in 2001 for 2002.

One may argue that utilities had different initial positions at the beginning of 2001 as a result of previous banking and endowments. Because a market exists for SO_2 allowances, this is not a problem. Utilities may purchase or sell at the market-clearing price the number of permits corresponding to their risk preferences.[18] Furthermore, banking of permits may be motivated by an

absolute obligation to supply, even if allowance prices are very high. Such a supply constraint is not present in our model because of the relatively low share of the permit price in the total production cost – namely, less than 3% of the total cost can be attributed to emissions permits (Considine & Larson, 2004).

5.2 Uncertainty

The difficulty here is to find a satisfactory measure of risk.[19] As stated by Lusardi (1998), "One needs to identify some observable and exogenous sources of risk that vary significantly across population."

We consider two sources of risk in this paper. First, we distinguish between states where restructuring is active, and states where it is not. Naturally, some utilities generate power for different states, which may not belong to the same type. In this case, we retain the main state where power is generated. This characteristic is specified through dummy variables D_{jk}, with $k = 1, 2$.

The second source of risk considered here comes from the intuition that generators with a higher share of coal-based power are more exposed under Title IV. These generators have a reduced ability to diversify their input if permit prices tend to increase. A utility producing exclusively with coal is fully exposed. The variable *coal*, representing the share of coal-based generation, is calculated for each utility. This last variable is corrected with a factor of emissions rate *pollut*.

5.3 Estimation

Because of the relatively small number of utilities considered, we only retain three characteristics for each utility: *source* is the total volume of power before any sale;[20] *revenue* gives the total revenue of the utilities; and D_{jk}, with $j = 1, 2, 3$, specifies

---

[18]Subject to their liquidity constraint.

[19]Contrary to the saving theory, the so-called self-selection bias, a critique addressed to Skinner (1988), is not present in our model. Indeed, because deregulation is a posterior fact, utilities do not select states where restructuring is or is not active in accordance with their risk preferences.

[20]The variable *source* is the sum of power generated and power sold for resale.

Table 2 Estimates

Variable	Means	Estimates	Student stat.
coal/pollut	0.147	0.661	2.165*
log(*source*)	6.950	−0, 141	−1.767**
revenue	1,324,061	8.40E-08	1.43
cooperative in deregulated market		2.320	1.848*
private in deregulated market		2.364	1.880*
public in deregulated market		2.935	2.499*
cooperative in regulated market		2.762	2.215*
private in regulated market		2.279	1.795*
public in regulated market		2.724	2.356*
Adjusted R^2		0.195	
Number of observations	137	137	

* significant at $P \leq 0.10$ level
** significant at $P \leq 0.05$ level

the type of owner, namely, cooperative, private, or public.

Following Kazarosian (1997), we perform the following semi-log regression:

$$ratio_i = \frac{coal_i}{pollut_i} + log(source_i)$$

$$+ revenue_i + \sum_{j=1}^{3} \sum_{k=1}^{2} D_{i,jk}. \qquad (6)$$

The results are shown in Table 2, which gives estimates with Student statistics.

Except for *revenue*, the estimates are significant. We obtain six different categories considering each owner type in both regulation and deregulation cases.

5.4 Findings

The evidence indicates a small but significant effect of uncertainty on banking behavior.[21] The dummy coefficients are not significantly different in states where restructuring is active and in states where it is not for private and public owners, but

they are different for cooperative owners. However, considering only restructuring dummies, we observe different behaviors in regulated and deregulated states. Utilities hold fewer permits in deregulated states, perhaps providing support for prudence in the Kimball (1990) sense. However, the significantly positive coefficient on *coal/pollut* suggests imprudence, because the more exposed the utility is, the more allowances it banks. As this coefficient is larger in absolute value, compared with the difference between coefficients in restructuring states and non-restructuring states, we may argue in favor of imprudence. So it appears that utilities would favor higher profits despite the resulting riskier probability distribution.

Concerning characteristics, because the coefficient on *revenue* is not significant, and because the one on *log(source)* is slightly positive, there does not seem to be any scale effect. Surprisingly, large and small utilities do not have significantly different approaches for banking.

6 Conclusion

The banking behavior of risk-averse firms has not previously been taken into account, theoretically or empirically, in the literature. This study fills this gap concerning emissions trading by providing a portfolio management approach to emissions permits. In this way, we draw attention to the financial aspect instead of the classical investment

---

[21] The adjusted R^2 of 0.195 is low, but its level is not surprising for cross-section estimation.

aspect, which in practice is generally limited to short-term analysis.[22]

From the viewpoint of economic policy, our results mean that regulators should consider the question of reducing permit price uncertainties by judicious choices regarding the design of the allowances market. In particular, we believe that the regulator may be able to improve the performance of the permits market by trading proactively in the allowances market and by allowing permit borrowing in a soft way. More precisely, the regulator can affect liquidity and reduce market price volatility by withholding or selling allowances to ensure that the market has the opportunity to function smoothly. The idea of possible welfare gains from governmental intervention is unfortunately not implemented in practice, although this policy recommendation is not new (Baldursson & von der Fehr, 2004; Dales, 1968). With regard to permit borrowing, it is well known that, in theory, emissions trading is efficient between periods only if the banking and borrowing of allowances are permitted (Rubin, 1996). However, the permitted use of allowances from a future period for compliance during the current period,[23] creates a fairly evident risk for the environment, because a firm that uses borrowed allowances during a given period may cease operation before the borrowed allowances are repaid through lower emissions. Moreover, one can imagine that firms make no abatement efforts voluntarily, borrowing permits and lobbying at the end of the program for a less drastic cap. For these reasons, unlimited borrowing of permits is not allowed in practice. However, the European emissions trading scheme for carbon dioxide, which started in 2005, allows a soft way for permit borrowing that should be generalized in other markets. This rule gives firms permission to use the $t + 1$ initial allocation to comply with the commitment period t. In this way, uncertainty is reduced and risk-averse firms should be less reluctant to sell permits compared with the case where only banking is allowed.

---

[22]For instance, a scrubber needs 2 or 3 years to be built.

[23]With the implicit commitment that repayment will be made in the form of equivalent reductions in a future period.

Appendix

The 137 utilities concerned with the present study are:

Alabama Electric Coop Inc, Alabama Power Co, City of Ames, Appalachian Power Co, Arizona Electric Pwr Coop Inc, Arizona Public Service Co, Arkansas Electric Coop Corp, Associated Electric Coop Inc, Atlantic City Electric Co, City of Austin, Black Hills Power & Light, Carolina Power & Light Co, City of Cedar Falls, Central Electric Power Coop, Central Illinois Light Co, Central Iowa Power Coop, Cincinnati Gas & Electric Co, CLECO Power LLC, City of Colorado Springs, City of Columbia, Columbus Southern Power Co, Consolidated Edison Co-NY Inc, Consumers Energy Co, Corn Belt Power Coop, Dairyland Power Coop, Dayton Power & Light Co, Deseret Generation & Tran Coop, Detroit Edison Co, Dominion Virginia Power, City of Dover, Duke Power Co, East Kentucky Power Coop Inc, Electric Energy Inc, Entergy Arkansas Inc, Entergy Gulf States Inc, Entergy Louisiana Inc, Entergy Mississippi Inc, Entergy New Orleans Inc, Florida Power & Light Co, Florida Power Corp, City of Fremont, Gainesville Regional Utilities, Georgia Power Co, City of Grand Island, Grand River Dam Authority, Great River Energy, Gulf Power Co, City of Hastings, Henderson City Utility Comm, City of Holland, Holyoke Water Power Co, Hoosier Energy R E C Inc, The Illuminating Co, City of Independence, Indiana Michigan Power Co, Indiana-Kentucky Electric Corp, Indianapolis Power & Light Co, Jacksonville Electric Auth, City of Jamestown, City of Kansas City, Kansas City Power & Light Co, Kentucky Power Co, Kentucky Utilities Co, KeySpan Generation LLC, City of Lake Worth, City of Lakeland City of, Lansing, City of Los Angeles, Louisville Gas & Electric Co, Lower Colorado River Authority, Madison Gas & Electric Co, Manitowoc Public Utilities, City of Marquette, MDU Resources Group, Inc, Michigan South Central Pwr Agy, MidAmerican Energy Co, Minnesota Power Inc, Minnkota Power Coop Inc, Mississippi Power Co, Monongahela Power Co, City of Muscatine, Nebraska Public Power District, Nevada Power Co, Northern Indiana Pub

Serv Co, Northern States Power Co, Ohio Power Co, Ohio Valley Electric Corp, Oklahoma Gas & Electric Co, Omaha Public Power District, Orlando Utilities Comm, Otter Tail Power Co, City of Owensboro, Pacific Gas & Electric Co, City of Pella, Pennsylvania Power Co, Platte River Power Authority, Portland General Electric Co, Power Authority of State of NY, PSI Energy Inc, Public Service Co of Colorado, Public Service Co of New Hampshire, Public Service Co of Oklahoma, City of Richmond, Rochester Gas & Electric Corp, Rochester Public Utilities, Salt River Proj Ag I & P Dist, San Antonio Public Service Bd, San Miguel Electric Coop Inc, Savannah Electric & Power Co, Seminole Electric Coop Inc, Sempra Energy Resources, Sierra Pacific Power Co, City of Sikeston, South Carolina Electric & Gas Co, South Carolina Genertg Co Inc, South Carolina Pub Serv Auth, South Mississippi El Pwr Assn, Southern California Edison Co, Southern Illinois Power Coop, Southwestern Electric Power Co, Southwestern Public Service Co, City of Springfield, Sunflower Electric Power Corp, City of Tallahassee, Tampa Electric Co, City of Taunton, Tennessee Valley Authority, Texas Municipal Power Agency, Toledo Edison Co, Tri-State G & T Assn Inc, Tucson Electric Power Co, Vectren Energy Delivery, WE Energies, Westar Energy, Western Farmers Elec Coop Inc, Wisconsin Public Service Corp, Wyandotte Municipal Serv Comm.

Acknowledgements The authors sincerely acknowledge Professor Louis Eeckhoudt from Catholic University of Mons and Professor Jacques Percebois from University of Montpellier for fruitful discussion on an earlier version of this paper.

References

Albrecht, J., Verbeke T., & de Clercq, M. (2004). Informational Efficiency of the U.S. SO₂ Permit Market, Faculty of Economics and Business Administration, Working Paper No. 250, Ghent University, Belgium.

Arimura, T. H. (2002). An empirical study of the SO₂ allowance market: effects of PUC regulations, *Journal of Environmental Economics and Management, 44,* 271–289.

Bailey, E. M. (1998). Allowance trading activity and state regulatory rulings: evidence from the u.s. acid rain program, MIT CEEPR Working Paper 98006, Center for Energy and Environmental Policy Research, Massachussetts Institute of Technology, Cambridge, MA.

Baldursson, F. M., & von der Fehr, N.-H. M. (2004). Price volatility and risk exposure: on market-based environmental policy instruments, *Journal of Environmental Economics and Management, 48,* 682–704.

Ben-David, S., Brookshire, D., Burness, S., McKee, M., & Schmidt, C. (2000). Attitudes toward risk and compliance in emission permit market, *Land Economics, 76,* 590–600.

Bohi, D. R., & Burtraw, D. (1997). SO₂ Allowance trading: how do expectations and experience measure up? *Electricity Journal, 10,* 67–75.

Browning, M., & Lusardi, A. (1996). Household saving: micro theories and micro facts, *Journal of Economic Literature, 34,* 1797–1855.

Burtraw, D. (1996). The SO2 emissions trading program: cost savings without allowance trades, *Contemporary Economic Policy, 14,* 79–94.

Carlson, C., Burtraw, D., Cropper, M., & Palmer, K. L. (2000). Sulfur dioxide control by electric utilities: what are the gains from trade, *Journal of Political Economy, 108,* 1292–1326.

Carlson, D. A., & Sholtz, A. M. (1994). Designing pollution market instruments: a case of uncertainty, *Contemporary Economic Policy, 12,* 114–125.

Cason, T., & Gangadharan, L. (2003). Transactions costs in tradable permit markets: an experimental study of pollution market designs, *Journal of Regulatory Economics, 23,* 145–165.

Chicago Climate Exchange. (2004). *The Sulfur Dioxide Emission Allowance Trading Program: Market Architecture, Market Dynamics and Pricing*, Chicago Climate Exchange, Inc., Chicagor, IL.

Considine, T. J., & Larson, D. F. (2004). The Environment as a Factor of Production, World Bank Policy Research Working Paper WPS 3271, Washington DC.

Cronshaw, M., & Kruse, J. B. (1996). Regulated firms in pollution permit markets with banking, *Journal of Regulatory Economics, 9,* 179–189.

Cropper, M. L., & Oates, W. E. (1992). Environmental economics: a survey, *Journal of Economic Literature, 30,* 675–740.

Dales, J. H. (1968). *Pollution, Property and Prices*, University of Toronto Press, Toronto, Canada.

Eeckhoudt, L., & Gollier, C. (2005). The impact of prudence on optimal prevention, *Economic Theory, 26,* 989–994.

Ellerman, D. A., & Montero, J. P. (1998). The declining trend in sulfur dioxide emissions: implications for allowances prices, *Journal of Environmental Economics and Management, 36 ,* 26–45.

Ellerman, D. A., Joskow, P. L., Schmalensee, R., Montero, J. P., & Bailey, E. M. (2000). *Markets for Clean Air: The U.S. Acid Rain Program*, Cambridge University Press, New York.

Energy Information Administration. (2004). *Annual Energy Review 2003*, US Department of Energy, EIA, Washington DC.

Godby, R. W., Mestelman, S. R., Muller, R. A., & Welland, J. D. (1997). Emissions trading with shares and coupons when control over discharges is uncertain, *Journal of Environmental Economics and Management, 32,* 359–381.

Gollier, C. (2001). *The Economics of Risk and Time,* MIT Press, Cambridge, MA.

Hahn, R. W. (1984). Market power and transferable property rights, *Quarterly Journal of Economics, 99,* 753–765.

Hahn, R. W., & May, C. A. (1994). The behavior of the allowance market: theory and evidence, *Electricity Journal, 7,* 28–37.

Hennessy, D. A., & Roosen, J. (1999). Stochastic pollution, permits and merger incentives, *Journal of Environmental Economics and Management, 37,* 211–232.

Holthausen, D. M. (1979). Hedging and the competitive firm under price uncertainty, *American Economic Review, 69,* 989–995.

Joskow, P. L., & Schmalensee, R. (1998). The political economy of market-based environmental policy: the u.s. acid rain program, *Journal of Law and Economics, 41,* 89–135.

Joskow, P. L., Schmalensee, R., & Bailey E. M. (1998). The market for sulfur dioxide emissions, *American Economic Review, 88,* 669–685.

Kazarosian, M. (1997). Precautionary savings – a panel study, *Review of Economics and Statistics, 79,* 241–247.

Keeler, A. G. (1991). Noncompliant firms in transferable discharge permit markets: some extensions, *Journal of Environmental Economics and Management, 21,* 180–189.

Kimball, M. S. (1990). Precautionary savings in the small and in the large, *Econometrica, 58,* 53–73.

Kling, C. & Rubin, J. (1997). Bankable permits for the control of environmental pollution, *Journal of Public Economics, 64,* 99–113.

Liski, M., & Montero J. P. (2005). A note on market power in an emission permits market with banking, *Environmental and Resource Economics, 31,* 159–173.

Lusardi, A. (1998). On the importance of the precautionary saving motive, *American Economic Review, Papers and Proceedings, 88,* 449–453.

Malik, A. S. (1990). Markets for pollution control when firms are non-compliant, *Journal of Environmental Economics and Management, 18,* 97–106.

Malik, A. S. (2002). Further results on permit markets with market power and cheating, *Journal of Environmental Economics and Management, 44,* 371–390.

Misiolek, W. S., & Elder, H. W. (1989). Exclusionary manipulation of markets for pollution rights, *Journal of Environmental Economics and Management, 16,* 156–166.

Montero, J. P. (1997). Marketable pollution permits with uncertainty and transactions costs, *Resource and Energy Economics, 20,* 27–50.

Montgomery, W. D. (1972). Markets in licenses and efficient pollution control programs, *Journal of Economic Theory, 5,* 395–418.

Rubin, J. (1996). A model of intertemporal emission trading, banking and borrowing, *Journal of Environmental Economics and Management, 31,* 122–136.

Sandmo, A. (1971). On the theory of the competitive firm under price uncertainty, *American Economic Review, 61,* 65–73.

Schennach, S. M. (2000). The economics of pollution permit banking in the context of title iv of the 1990 clean air act amendments, *Journal of Environmental Economics and Management, 40,* 189–210.

Schmalensee, R., Joskow, P. L., Ellerman, D. A., Montero, J. P., & Bailey, E. M. (1998). An interim evaluation of sulfur dioxide emissions trading, *Journal of Economic Perspectives, 12,* 53–68.

Skinner, J. (1988). Risky income, life-cycle consumption and precautionary savings, *Journal of Monetary Economics, 22,* 237–255.

Swift, B. (2001). How environmental laws work: an analysis of the utility sectors response to regulation of nitrogen oxides and sulfur dioxide under the clean air act, *Tulane Environmental Law Journal, 14,* 309–424.

Tietenberg, T. H. (1985). Emissions Trading: An Exercise in Reforming Pollution Policy, Resources for the Future, Washington DC.

van Egteren, H., & Weber, M. (1996). Marketable permits, market power, and cheating, *Journal of Environmental Economics and Management, 30,* 161–173.

Water Air Soil Pollut: Focus (2007) 7:573–579
DOI 10.1007/s11267-006-9110-x

Tradable Permit Systems: Considering Uncertainty in Emission Estimates

Paweł Bartoszczuk · Joanna Horabik

Received: 24 May 2005 / Accepted: 28 August 2006 / Published online: 25 January 2007
© Springer Science + Business Media B.V. 2007

Abstract We simulate the market for emission permits by considering uncertainty in emission inventory reports. The approach taken in this analysis is to enhance the emissions reported in each region by a certain part of their uncertainty when compliance with the Kyoto targets is being proved. While this formulation is not new in the literature, we define the uncertainty component in a way that enables comparison with the approach of effective permits presented in Nahorski, Horabik, and Jonas (2007) Compliance and emissions trading under the Kyoto protocol: Rules for uncertain inventories, (this issue). We show and explain that the transformation to effective permits bears additional costs apart from those resulting from shifting the Kyoto targets.

Keywords emissions trading · Kyoto Protocol · inventory uncertainty · effective permits

1 Introduction

The system for tradable emission permits has been designed as a cost-effective method of reducing emissions to the desired level. The permit trade, which results in an equalization of marginal abatement costs among pollution sources. In general, the literature provides strong support for the use of such a system as part of environmental policy (Ellerman & Decaux, 1998; Ellerman, Jacoby, & Decaux, 1998; Godal, 2000; Field & Field, 2002; Hill & Kriström, 2002; Holtsmark & Maestad, 2002; Sterner, 2003; Tietenberg, 1998). Permit systems for tradable emissions have been the most frequently used market-based instruments of environmental policy over the last decade.

Ellerman and Decaux (1998) applied marginal abatement cost (MAC) curves, which were generated using the Emissions Prediction and Policy Assessment (EPPA) model of the Massachusetts Institute of Technology (MIT) (Yang, Eckaus, Ellerman, & Jacoby, 1996), a recursive-dynamic multiregional computable general equilibrium (CGE) model. EPPA simulates projections of economic growth with the objective of producing scenarios of greenhouse gases (GHGs) and their precursors emitted as a result of simulated human activities and the real emission reductions that it would be possible to make (Babiker et al., 2001; Paltsev et al., 2005).

Countries listed in Annex B to the Kyoto Protocol have agreed to reduce their emissions to below base-year levels during the period 2008–2012 (Holtsmark & Maestad, 2002). Emissions trading among Parties to the Kyoto Protocol will start in 2008. Although Parties are responsible for estimating and reporting uncertainties in their emission estimates to the United Nations Framework Convention on Climate Change (UNFCCC), the Marrakesh Accords, which prepare for the effective participation of developing countries in the Kyoto Protocol process, did not set any bounds

P. Bartoszczuk (✉) · J. Horabik
Systems Research Institute, Polish Academy of Sciences,
Newelska 6, Warsaw 01-447, Poland
e-mail: Pawel.Bartoszczuk@ibspan.waw.pl

Springer

for uncertainty in tradable emissions (Monni, Syri, Pipatti, & Savolainen, 2007). We have no evidence that the emission targets set by the Protocol are sufficient, either for an individual country or for the world as a whole (Tietenberg, 1985).

Winiwarter (2007) points out that emission reductions as proposed by the signatories to the Kyoto Protocol are far too small to decrease increasing GHG concentrations and that new targets for future emissions should thus be established. Because of possible underreporting, a periodic review of the emission reduction targets should consider inventory uncertainty. The problem of uncertainty was extensively discussed by many authors at the 2004 workshop, "Uncertainty in Greenhouse Gas Inventories", held in Warsaw, Poland. Gillenwater, Sussman, and Cohen (2007) claim that the investigation of inventory uncertainty can make data more transparent and of higher quality. Monni et al. (2007) point out that differences among uncertainties in different emissions trading schemes can be substantial. Uncertainty varies among Parties to the Kyoto Protocol and among activities that generate emissions. The estimates gathered in Nahorski et al. (2007) show that uncertainty in greenhouse gas inventories has been estimated to be in the range of 5–20%, depending on the methodology used. Even if some computations need to be recalculated, it is believed that uncertainty may still be at least 12% (Winiwarter, 2007).

According to Nilsson, Shvidenko, and Jonas (2007) uncertainty can be defined as "an imperfection in knowledge of the true value of a particular parameter or its real variability in an individual or a group". It can be represented by a range of values calculated by various models or by qualitative measures. The Intergovernmental Panel on Climate Change (IPCC) has provided general guidance for uncertainty management in greenhouse gas emissions (IPCC, 2000). It underlines that the IPCC Tier-1 methodology relies on three points: (1) all individual emission sources are independent from each other; (2) the emissions show normal (Gaussian) distributions; and (3) uncertainties for greenhouse gases are smaller than 60%. Many uncertainty estimates are, in the final analysis, based on expert judgment and are thus very subjective. The calculated uncertainty in the total carbon dioxide (CO_2) emission trends for the years 1990–2002 in the Netherlands is ±3% (Brandes, Olivier, & van Oorschot, 2004). So far, official national estimates of other EU member states are reported to range below (see EEA, 2006, Table 1; but see also Table 1 in Nahorski and Jęda (2007). These uncertainties are shown by Rousse & Sévi (2004) and by others to have a measurable impact on behavior in emissions trading markets. They underline that uncertainty has a perturbatory effect on trading.

Nahorski et al. (2007) present a solution for tackling the problem of inventory uncertainty both under Kyoto verification conditions and under emissions trading. They introduce the risk that the real (but unknown) emissions actually exceed the reported levels because of inventory uncertainities. The authors propose that the compliance level should be shifted down by some fraction of the uncertainty level in order to prove compliance with the Kyoto target (the concept of "undershooting"). Thus, the emission reduction should undershoot the level of uncertainty, or its fraction, if we agree to bear some risk. Moreover, Nahorski et al. (2007) assume that the uncertainty of the purchased emissions contributes to the overall buyer's uncertainty and to the introduction of so-called effective permits – each party transforms its regular permits into effective permits and the transformation ratio depends on the reported inventory

Table 1 Initial emissions, changes in emissions, inventory uncertainty of carbon dioxide in different regions, Kyoto obligation; no risk of underreporting included in initial emissions

Region	Base year emissions (MtC/y)	Kyoto target (%)	Inventory uncertainty (%)	Marginal costs of abatement ($/tC)	Total cost (MUS$)
USA	1,345	7.0	13	313.7	89,343
OECDE	934	7.9	10	322.7	28,652
JAPAN	274	6.0	15	453.8	21,077
CANZ	217	0.7	20	216.5	10,477
EEFSU	1,337	1.7	30	0	0
TOTAL	4,107				149,549

uncertainty. Trading takes place in the effective permits market.[1]

In our research, effective permits are not considered. Instead, we apply only the concept of undershooting to an analysis of sensitivity in a permit market. We aim to show that, by avoiding the additional costs that effective permits require, modifications of the Kyoto verification condition to address inventory uncertainty result in a more efficient market solution.

Godal, Ermoliev, Klaassen, and Obersteiner (2003) also consider the problem of inventory uncertainty in emissions trading, describing a dynamic trading scheme under which regions approach a decision-making problem in two steps. First, for a given amount of permits, each region must spend resources either on abating emissions or on investing in monitoring (reducing inventory uncertainty). Permits are exchanged bilaterally between parties in the second optimization step. In our approach, we do not consider uncertainty reduction costs. Although taking uncertainty reduction costs into account is highly justifiable, assessing those costs is an extremely difficult task. Data on this kind of costs are virtually unavailable, as Godal et al. (2003) admit by modeling them in a simplified way.

2 The Modeling Framework

Our modeling exercise is aimed at analyzing the influence on market performance of incorporating effective permits. To do this we need to adopt the same rules of verification (i.e., undershooting) but without including effective permits. Effective permits merely solve the problem of the differing permit quality caused by imperfect inventories. The costs of this solution also need to be examined.

To do this, we incorporate the uncertainty component into the standard permit trade described by Tietenberg (1985) (i.e., when checking compliance with the Kyoto targets, we require each Party's emissions to be increased to account for the existence of uncertainty). In our model, the compliance-proving condition for each party is as formulated in Nahorski et al. (2007, equation 10). However, as these authors were consid-

ering effective permits, they had not only to transform both the emissions at the commitment year and the Kyoto targets into effective permit equivalents in Nahorski et al. (2007, equation 25), but also to use this modified condition in their permit trade calculations in Nahorski et al. (2007, equation 35). In that setting, the cost functions also had to be expressed in terms of effective permits in Nahorski et al. (2007, equation 33).

Before we state the optimization problem, we introduce the notation. We consider n regions with emission levels x_i ($i=1, 2,...,n$). The total costs of holding emissions in region i down to x_i is denoted by $C_i(x_i)$ (abatement cost function). We assume that cost functions $C_i(x_i)$ are positive, decreasing, and continuously differentiable for each region. The Kyoto target for each region i is indicated by K_i. The number of emission permits acquired by source is expressed by y_i (y_i is negative if region i is a net supplier of permits).

The problem is then formulated as follows:

$$\min_{x_i}[\sum_{i=1}^{n} c_i(x_i)] \tag{1}$$

subject to

$$x_i + (1 - \varsigma)(1 - 2\alpha)d_i n_i \leq K_i + y_1,$$
$$\text{for } i = 1, 2, \ldots, n \tag{2}$$

$$\sum_i y_i = 0 \tag{3}$$

The task is to minimize the sum of total emission reduction costs for n regions. There are $n+1$ side conditions. Equation 2 states that, for each of n regions, the reported emissions plus the potentially unreported emissions, that is, $x_i + (1 - \varsigma)(1 - 2\alpha) d_i n_i$, must be below (or equal to) the Kyoto target K_i modified by the amount of bought/sold permits y_i. Finally, the sum of permits bought and sold equals 0 in order to maintain market equilibrium (Eq. 3).

To evaluate the comparison we need to consider uncertainty (undershooting) closely following Nahorski et al. (2007). In the uncertainty belt, expressed as $(1 - \varsigma)(1 - 2\alpha)d_i n_i$, the parameter α represents the risk of not satisfying the Kyoto target because of the uncertainty in the inventory estimate $\alpha \in [0; 0.5]$; and it is set so as to be common for all participants. The expression $(1-2\alpha)$ allows for adjustment in the extent to which uncertainty is included in the scheme. The relative

---

[1] More information at http://www.ibspan.waw.pl/GHGUncert2004/ prezent/Nahorski_prezentacja.pdf.

Springer

uncertainty of emission reports is represented by d_i. As those uncertainties vary for different regions or countries (Winiwarter, 2007), they are also indexed by i. Base-year (1990) emissions at source i are depicted by n_i, while x_i represents emissions in the commitment year (2010). It should be emphasized that the study takes the year 2010 as representative of the first commitment period, which includes the years 2008 through 2012. Finally, in order to account for dependence between uncertainty in the base and commitment years, we also apply parameter ς. As, in the literature, the dependence is estimated for 0.65–0.7, we adopt $\varsigma= 0.7$, which means that parties are penalized only with respect to 30% of their uncertainty records.

3 The Data

For comparability of results with Nahorski et al. (2007), we make use of the same data set. The data should be regarded as illustrative because of its high level of aggregation and rough uncertainty estimations. Countries are aggregated into five groups: United States (US); countries of the Organisation of Economic Cooperation and Development, Europe (OECDE); Japan, Canada/Australia/New Zealand (CANZ); and the countries of Eastern Europe/former Soviet Union (EEFSU). In our approach, to avoid complicated calculations and because of the scarcity of data, we do not consider the stochastic model of uncertainty. We assume arbitrarily that the uncertainty coefficient d_i equals a given percentage of reported emissions in the base year (Godal et al., 2003).

The carbon emission reduction constraints used for this study are based on the commitments made by countries to the Kyoto Protocol. Five of the six regions belonging to Annex B are obliged to reduce emissions. However, it is predicted that the FSU emissions will be below the level that this nation committed to under Kyoto. The difference between the FSU commitment and expected emissions is often described as "hot air" and can be sold as a "right to emit".

4 Results of Simulations

Below we present the results of trading when undershooting is incorporated without effective permits. As inventory uncertainty is present for every single party, transformation into trade with effective permits

requires higher emission reductions. Our proposition is expected to result in total emissions (for all parties) higher than in Nahorski et al. (2007) and lower compared with the base case under which no uncertainty is taken into account. Costs in our solution will be lower than in the case of effective permits covered by Nahorski et al. (2007).

To portray the above statements and show the scale of difference between the approaches, the numerical example for the available aggregated data is given below.

The regions begin to trade, and the market permit price settles when marginal costs equalize among participants.

With uncertainty disregarded ($\alpha=0.5$), the results are obviously exactly the same as in Nahorski et al.

Table 2 Results of simulation for different levels of risk α; trading with permits according to the concept of undershooting without effective permits

Region	Reported emissions (MtC/y)	Market price of permits ($/tC)	Total cost (MUS$)	Permits traded (MtC/y)
Risk parameter $\alpha=0.5$				
USA	1,561.6	142.5	18,433	310.8
OECDE	959.4	142.5	5,602	99.1
JAPAN	321.1	142.5	2,059	63.5
CANZ	248.4	142.5	4,583	32.9
EEFSU	807.8	142.5	6,473	−506.3
	3,898.3		37,150	0.0
$\alpha=0.3$				
USA	1,511.0	170.4	26,352	281.0
OECDE	944.0	170.4	8,010	106.4
JAPAN	315.5	170.4	2,943	62.4
CANZ	235.8	170.4	6,549	26.0
EEFSU	790.0	170.4	9,245	−475.8
	3,796.3		53,101	0.0
$\alpha=0.1$				
USA	1,460.6	198.2	35,649	251.5
OECDE	928.7	198.2	10,835	113.5
JAPAN	309.8	198.2	3,982	61.7
CANZ	223.3	198.2	8,861	18.7
EEFSU	772.3	198.2	12,508	−445.4
	3,694.7		71,838	0.0
$\alpha=0$				
USA	1,435.4	212.1	4,824	236.8
OECDE	921.0	212.1	12,408	117.0
JAPAN	307.0	212.1	4,560	61.3
CANZ	217.0	212.1	10,147	15.0
EEFSU	763.5	212.1	14,325	−430.2
	3,643.8		82,266	0.0

(2007). In the scenario with $\alpha=0.3$, when a risk of 30% is taken that each Party's actual emissions can be above the Kyoto obligation, the results are as follows. The total reported emissions (emissions aggregated for all the participants) are 3,796 MtC/y, which is slightly (2.3%) higher than the ineffective permit scenario (see Table III in Nahorski et al., 2007). However, total costs ($US53.101 million) are 23% lower than in the effective permit case. The market permit price is 170.4 $/tC, compared with the price of effective permits 198.4 $/tC (permit prices measured in reported emissions vary between 191.2 $/tC and 196 $/tC for $\alpha=0.3$). Because of the lower price in our case, permit turnover is higher. One further aspect of emissions trading can be analyzed from the table: the fact that benefits from emissions trading are not evenly distributed among participants. As stated in Ellerman and Decaux (1998), regions with autarkic marginal cost (marginal cost without trade) further from the trading equilibrium will benefit more than those regions with autarkic marginal cost closer to the trading equilibrium. In our analysis, the greatest benefits are obtained by (1) Japan, which imports the most permits and which, without trading, would have to bear a marginal cost of 453.8 $/tC; and (2) the FSU, which provides all the permits on the market at zero cost. For a detailed analysis of the distribution of benefits from trade (although evaluated for another data set), see Bartoszczuk (2004).

Regarding the sensitivity of the results when we change the degree of included uncertainty, Table 2 also shows the scenarios with risk parameter α equal to 0.1 and 0. Quite naturally when we diminish α (i.e., account for more uncertainty), a lower emission level is required by the Kyoto Protocol which, in turn, results in increased costs.

Finally, in the scenario with risk parameter $\alpha=0$ (i.e., when the entire uncertainty belt is taken into account), the total amount of emissions allowed is about 6% higher than in the effective permit scenario and the total costs are 38% lower. The amount of permits traded on the market is also 38% higher.

5 Conclusions

Generally, our approach of adopting the undershooting rule without effective permits is an intermediate solution to the problem of inventory uncertainty within emissions trading. It is more cost-effective than the effective permit case, but it is also naturally more expensive than the nonuncertainty scenario of $\alpha=0.5$. Thus, the rise in costs in the approach taken by Nahorski et al. (2007) can be broken into two parts: that resulting from undershooting conditions and that emerging from transition into effective permits. For the considered data set, the undershooting condition was responsible for 46–49% of the increase in costs. The ratio remains about the same when we change α.

In practicality, applying the effective permits approach requires additional agreements among participants. While such agreements are difficult to attain, our analysis could be informative for decision makers as it sheds light on another aspect of the design of permit trading.

Appendix

Emissions Trading and Marginal Abatement Cost Curves

Below we present the economic bases of the emissions trading mechanism. It introduces a new type of property right that allows a specified amount of

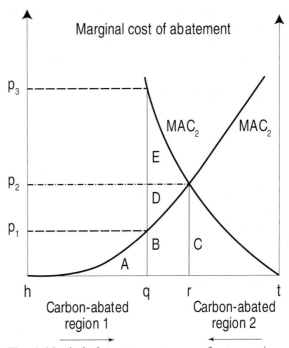

Fig. 1 Marginal abatements cost curves for two regions: MAC_1 and MAC_2

pollutant to be emitted. Thus, the total number of permits held by all sources puts a limit on the total quantity of emissions. Permits can be sold to anyone participating in the permit market. The system is initialized by central decision makers who decide on the number of permits to be put into circulation. As the total number of permits is usually lower than current total emissions, some emitters will receive fewer permits than their current emissions would normally allow.

Regional purchase or sale of permits until their marginal costs are equalized:

$$MAC_1 = MAC_2 = P_2$$

This way the aggregate emission reduction is reached at the least cost for the whole market. The difference between the market price and the marginal cost in the absence of trade creates a potential gain that is shared between the two trading regions. This is illustrated in Fig. 1.

The origin of the marginal cost of control for the first source (MAC_1) is the left-hand axis and the origin of the marginal cost of control for the second source (MAC_2) is the right-hand axis. The diagram represents all possible allocations of the reduction between the two sources. The left-hand axis represents an allocation of the entire control responsibility to the second region, while the right-hand axis represents a situation in which the first source bears responsibility (Tietenberg, 1985). Initially region "1" must reduce hq amount of pollution, while region "2" tq amounts of pollution (looking from the right side of the axis). Total abatement cost is given by the area below the marginal abatement cost curves MAC_1 and MAC_2, respectively. Before trading, the total abatement cost for region "1" is the field "A", and for region "2" is the sum of the areas $B+C+D+E$. Region "2" buys permits to emit more than is allowed (in other words, it reduces only the amount (tr) instead of the amount (tq), while region "1" abates more than it is obliged to do (hr), simultaneously selling (qr) permits.

After exchange of permits, costs for both sources is represented by the area ($A+B+C$). The sum of the area A and B is the cost of control for the first source, while the area C is the cost of control for the second. The area $D+E$ represents the amount saved by emissions trading. The costs of emission reduction is

minimized, as the marginal costs are equalized across the emitters. Both regions have incentives to trade, as the marginal cost of control for the second region is higher than that for the first region. The second region will lower its costs as long as it can buy permits from the first region at a price lower than p_3. When the price equals p_2, neither region would have any further incentive to trade.

References

Babiker, M. H., Reilly, J. M., Mayer, M., Eckaus, R. S., Wing, I. S., & Hyman, R. C. (2001). The MIT emissions prediction and policy analysis (EPPA) model: Revisions, sensitivities, and comparisons of results. *MIT Report No. 70*, (Internet edition).

Bartoszczuk, P. (2004). Tradable emission permits as efficient strategy for achieving environmental goals. In *Proceedings of the international workshop on uncertainty in greenhouse gas inventories: Verification, compliance and trading*, held 24–25 September in Warsaw, Poland, pp. 143–150. Available at http://www.ibspan.waw.pl/ GHGUncert2004/papers.

Brandes, L. J., Olivier, J. G. J., & van Oorschot, M. M. P. (2004). Validation, verification and uncertainty assessment for the Netherlands. Emission inventory. In *Proceedings of the international workshop on uncertainty in greenhouse gas inventories: Verification, compliance and trading*, held 24–25 September in Warsaw, Poland, pp.19–33. Available at http://www.ibspan.waw.pl/ GHGUncert2004/papers.

EEA (2006). Annual European community greenhouse gas inventory 1990–2004 and inventory report 2006. Technical report no. 6, European Environment Agency, Copenhagen, Denmark. Available at: http://reports.eea.europa.eu/ technical_report_2006_6/en.

Ellerman, A. D., & Decaux, A. (1998). Analysis of post-Kyoto CO_2 emissions trading using marginal abatement curves, emission and ambient permits: A dynamic approach. *Report No. 40, MIT Joint program on the science and policy of global change*. Massachusetts Institute of Technology, Cambridge, MA, USA, 15, 39–56.

Ellerman, A. D., Jacoby, H. D., & Decaux, A. (1998). The effects on developing countries of the Kyoto protocol, and analysis of CO_2 emissions trading. *Report of the joint program on the science and policy of global change Massachusetts Institute of Technology*. Cambridge, MA, USA.

Field, B. C., & Field, M. K. (2002). *Environmental Economics, An Introduction*. New York: McGraw-Hill.

Gillenwater, M., Sussman, F., & Cohen, J. (2007). Practical applications of uncertainty analysis for national greenhouse gas inventories (this issue).

Godal, O. (2000). Simulating the carbon permit market with imperfect observations of emissions: Approaching equilibrium through sequential bilateral trade. *Interim Report IR-00-060*, International institute for applied systems analysis (IIASA), Laxenburg, Austria.

Godal, O., Ermoliev, Y., Klaassen, G., & Obersteiner, M. (2003). Carbon trading with imperfectly observable emisssions. *Environmental and Resource Economics, 25,* 151–169.

Hill, M., & Kriström, B. (2002). *Sectoral EU trading and other climate policy options: Impacts on the Swedish economy,* Stockholm School of Economics, Stockholm, Sweden. See http://www.sekon.slu.se/~bkr/hk2.pdf, 23.08.2004.

Holtsmark, B., Maestad, O. (2002). Emission trading under the Kyoto protocol-effects on fossil fuels markets under alternative regimes. *Energy Policy, 30,* 207–218.

IPCC (2000). *Good practice guidance and uncertainty management in national greenhouse gas inventories.* Japan: IPCC-TSU NGGIP.

Monni, S., Syri, S., Pipatti, & Savolainen, I. (2007). Extension of Eu emissions trading scheme to other sectors and gases: Consequences for uncertainty of total tradable amount (this issue).

Nahorski, Z., Jęda, W. (2007). Processing national CO_2 inventory emissions data and their total uncertainty estimates (this issue).

Nahorski, Z., Horabik, J., & Jonas M. (2007). Compliance and emissions trading under the Kyoto protocol: Rules for uncertain inventories (this issue).

Nilsson, S., Shvidenko, A., & Jonas, M. (2007). Uncertainties of the regional terrsetial biota full carbon account: A systems analysis (this issue).

Paltsev, S., Reilly, J. M., Jacoby, H. D., Richard, S. E., McFarland, J., Sarofirm, M., et al. (2005). *The MIT emissions prediction and policy analysis (EPPA) model: Version 4, Report No 125.* (Internet Edition).

Rousse, O., & Sévi, B. (2004). Portfolio managment of emission permits and prudence behaviour uncertainties of the regional terrestial biota full carbon account: A systems analysis. In *Proceedings of the international workshop on uncertainty in greenhouse gas inventories: Verification, compliance and trading,* held 24–25 September in Warsaw, Poland, pp. 135–142. Available at http://www.ibspan.waw.pl/ GHGUncert2004/papers.

Sterner, T. (2003). *Policy Instruments for Environmental and Natural Resource Management,Resources for the future.* Washington, D.C., USA.

Tietenberg, T. (1985). *Emissions Trading, An Exercise in Reforming Pollution Policy, Resources for the future.* Washington, D.C., USA.

Tietenberg, T. (1998). *Environmental economics and policy.* Reading, MA: Addison-Wesley.

Winiwarter, W. (2007). National greenhouse gas inventories: Understanding uncertainties versus potential for improving reliability (this issue).

Yang, Z., Eckaus, R. S., Ellerman, A. D., & Jacoby, H. D. (1996). The MIT emission prediction and policy analysis (EPPA) model. *Report No. 6,* MIT joint program on science and policy of global change, Cambridge, MA, USA.